Controlled Burn

By D. Dwayne Edwards

Copyright 2025

Controlled   **Burn**

Published by Curiosity Press
United States of America

Hardcover - ISBN 979-8-9943105-8-8
Paperback - ISBN 979-8-9943105-7-1
eBook - ISBN 979-8-9943105-6-4

Library of Congress Control Number: In Process

Author: D. Dwayne Edwards

First Edition

For my curious and amazing daughter, Mallory.

Prologue

Flour is not as innocent as it would like you to believe.

Sure, it looks like pale, harmless dust, and domestic. But pack enough of it in the proper ratio with oxygen, and you've got a low-yield explosive. A bakery, in purely chemical terms, is a battlefield of both starch and fire. Every loaf that emerges golden and fragrant from an oven is a carefully negotiated truce among volatile powders, runaway heat, and water molecules staging their own jailbreak into steam.

You don't think about that when you're eating a croissant. You're enjoying the butter, flakiness, and maybe the slight guilt of devouring three in one sitting. But inside that delicate lattice of browned crust and airy crumb is a battleground of chemical reactions, with gluten scaffolds stretching like tension cables on a bridge, sugars breaking down into caramel notes via Maillard reactions, and carbon dioxide inflating millions of microscopic balloons before bursting and leaving behind the cavernous interior we call 'crumb'.

Many of us already know that baking is not an art; it is a science. In fact, it's a subset called chemistry, and controlled, repeatable, measurable chemistry. The kind that doesn't forgive impatience or sloppy math. The kind that rewards those who can see not just dough but the molecules inside it; hydrogen bonds aligning, protein chains tangling, enzymes gnawing away at starch like tiny, efficient machines.

The baker is both a scientist and a sorcerer, wielding heat like Elon Musk with a flamethrower, and managing fermentation like a slow biological reactor.

Proofing isn't magic, it's microbial engineering. Mixing isn't just blending; it's orchestrating protein alignment at a microscopic level. Every motion, every pause, every degree on the thermometer is data.

This book is about that hidden battlefield. It is not a collection of recipes. Recipes are blueprints without physics. It is the operating manual for the chemistry set you already own: most of it is already in your kitchen. There, you'll find stoichiometry masquerading as measuring cups, thermodynamics hidden in oven dials, and biochemistry bubbling away in your starter jar.

So treat this like a novel in which the characters are carbon, nitrogen, hydrogen, and oxygen, locked in endless drama. Manage it like a lab manual where the experiments are delicious, and the results are edible.

Here's the simple truth: if you want to bake like an artist, you must first think like a chemist. And remember, flour may look innocent, but it's waiting to react.

Chapter One – Bread: Humanity's First Experiment

Bread is civilization disguised as food. We often think of bread as an accessory, something to hold sandwich fillings together, a sidekick to butter or soup. But bread was, arguably, our species' first real chemistry experiment.

Archaeologists like to tell the story this way: around 14,000 years ago, a group of hunter-gatherers in what is now Jordan ground wild grains into coarse meal, mixed it with water, and cooked it on hot stones. What came off the rocks was proto-bread, flat, dense, and chewy. By itself, that's not revolutionary.

Humans had been eating seeds, roots, and tubers long before that. But when those grains were mashed and heated, starch molecules unraveled, proteins tangled, and flavors deepened. In other words, raw grass seeds became digestible food. That was alchemy by the campfire.

A few thousand years later, another accident, fermentation, transformed this humble flatbread into something far stranger: bread that rose. The Egyptians get the historical credit.

Hieroglyphics dating back to 1500 BCE depict bakers kneading, shaping, and stacking loaves that look suspiciously like the baguettes still sold in Cairo markets today.

Somewhere between grinding and baking, the dough was left out long enough for wild yeasts and lactic acid bacteria to move in. They fed on sugars, released carbon dioxide, and filled the dough with tiny air pockets. For the first time, people bit into something soft, light, and almost otherworldly.

That leap, from hard flatbread to airy loaf, wasn't just culinary. It was biochemical. Bread became the first example of controlled use of microbial power in food. Long before Pasteur sketched the yeast cells he discovered under his microscope, ancient bakers conducted daily fermentation experiments, guided only by taste, smell, and trial and error.

The Physics of a Loaf

In essence, discard the flowery language, and you are left with the unadorned, somewhat stark truth: Bread is merely a structural arrangement, a collaborative effort among a triumvirate of key players: starch, protein, and a bustling metropolis of microbes.

Let us begin with starch, that ubiquitous yet often misunderstood character of the plant world. It's Mother Nature's pantry staple and a bit of a hoarder.

Imagine a plant: after a long day of soaking up the sun, it has all this extra glucose (*sugar*) lying around. Instead of a checking account, it uses starch as its savings account, a place to stash that energy for a rainy day, a night, or the long winter.

Chemically, it's a polysaccharide, a fancy, ten-dollar word (*plus twenty-seven cents for inflation*) for "many sugars stuck together." Think of it as a massive, intricate chain-link fence made entirely of glucose units. There are two types of "links" in this fence:

Our first contestant is Amylose, the straight-laced, linear one. It likes to coil into a helix, which is why adding iodine to a starchy food like a potato turns it deep blue-black. It's resistant to quick digestion, the strong, silent type.

Found in rice and seeds, Amylopectin joins the show. This one is a social butterfly. Highly branched, it's less about structure and more about volume, making up the bulk (*around 70-80%*) of the starch granule. Its branching allows it to break down more quickly, providing a quick energy burst.

When it comes to the kitchen, this stuff is the ultimate multitasker. Add some heat and water, and these granules swell up like they're at a buffet, turning into a sticky paste that thickens gravies and sauces. In its raw form, it's a hard, semi-crystalline powder. It gives bread and other baked goods body. When extracted, it's an odorless, tasteless, white powder, proving that even the most complex characters can have a bland, unassuming cover.

So, in short, starch is the plant world's strategic energy reserve: a complex, coiled, and branched carbohydrate that's simultaneously a humble powder, a powerful thickener, and the very foundation of your daily bread.

Every story, like every town, needs a lovable, hard-working blue-collar type. In our humble little story, that character is protein, the bouncer of the biological world! It's that dapper, hard-working molecule in your rye bread that decides who gets in, who gets out, and generally keeps the whole joint from falling apart.

Here's the skinny on this biological bricklayer. It's a long chain of amino acids strung together like a string of very particular pearls. There are twenty types, each with its own agenda, and the order on that string is crucial. A misplaced link is like putting the bouncer on kitchen duty and the chef on the door, trying to keep out the riffraff. Chaos ensues.

The Grand Design: This chain doesn't just lie there like a lazy snake; it folds into a very specific, often globular shape. This molecular origami is where the magic happens.

So what does it do? Everything, really. It's the muscle (*literally, in your own body*), the transport service (*hemoglobin carries oxygen*), the defense department (*antibodies*), and the tiny toolkit (*enzymes speed up reactions*). Without it, you're just a bag of water and misplaced intentions.

The it factor of rye, specifically its protein, is a free spirit compared to wheat's more rigid gluten. Wheat protein is a team player, forming a strong, elastic matrix that makes bread rise high. Rye protein, known as secalin, is more of a solo act, resulting in a denser, charmingly stout loaf with a distinctive, robust flavor – the kind of bread that knows its own mind and isn't trying to be something it's not. It's a reliable, if less airy, character.

So, next time you bite into a slice of dense, flavorful rye, tip your hat to the protein, the unsung, slightly stubborn backbone of the operation.

Ah, the miniature titans of fermentation! These aren't just tiny organisms; they're the invisible hand behind some of humanity's greatest hits: bread, beer, and a whole lot of *sour* attitudes.

Microbes are the world's most prolific, uninvited guests. They were here first, will be here last, and, frankly, they run the show while remaining too small for the naked eye to see. They are nature's original alchemists, turning the mundane into the magnificent (*or, occasionally, the repulsive*).

Some make us ill, while others ensure our gut biome isn't a wasteland. They're the ultimate multitaskers, and they don't even demand a salary, just a little sugar or a cozy place to call home.

Meet Saccharomyces cerevisiae, quite literally the life of the party! Better known as brewer's or baker's yeast, it's the overachiever of the microbial world. Its name literally means "sugar fungus of beer," which tells you everything you need to know about its priorities. This single-celled rockstar has been instrumental in human civilization for millennia, essentially turning sugars into alcohol and carbon dioxide.

In the bakery, it's the diva responsible for that delightful rise, puffing up dough with its CO_2 emissions (*a polite way of saying yeast flatulence*) to create the perfect loaf of bread. In the brewery, it's the boozy friend who converts all that nice barley sugar into ethanol, making happy hour possible.

It's the organism that single-handedly turns flat, sad grape juice or grain water into a party, and then has the spirit to become one of the most-studied model organisms in science. It's basically a tiny, budding celebrity with 1 billion subs and growing.

Now for Lactobacillus, the industrious, rod-shaped bacterium that skipped the sweet life and went straight for the sour. These are the culinary "Three Stooges" of the sour world, converting carbohydrates into lactic acid with impressive efficiency, aided by minimal slapstick.

It's the reason your yogurt has that tang, your cheese exists, and your sauerkraut isn't just plain cabbage.

In the human body, it's a key player in the gut and other systems, creating a protective acidic barrier that keeps the real bad guys at bay.

While Saccharomyces gets the credit for the buzz and the fluff, Lactobacillus is the quiet workhorse, contributing complex flavors to sourdough and keeping your digestive system in line. It's the health-conscious friend who always chooses the tart option.

Combine these three: starch, protein, and microbes, and you don't just get food. You get a system, a small-scale bioreactor humming on your countertop, producing both calories and culture.

Cooking breaks the vault open, releasing calories that would otherwise be locked in.

Bread and Civilization

Why does bread matter so much? Because once you figure out how to turn inedible seeds into caloric gold, you can stop chasing mammoths across the tundra. You can farm. You can settle. Cities form, hierarchies emerge, armies march, and behind them all, someone is baking.

The Sumerians wrote hymns to their beer god, Ninkasi, but they might as well have written them to bread. Both beer and bread rely on the same enzymatic trick: malted grains convert starch into sugar, which feeds yeast, which produces alcohol or gas. The twin inventions of bread and beer, fermented grains in solid and liquid form, were arguably the cornerstones of urban civilization.

Even the Romans, famous for their roads and aqueducts, measured social stability by grain supply. The phrase "bread and circuses" wasn't metaphorical. Grain shipments from Egypt sustained the empire; bakeries became a key part of its political infrastructure. Control the ovens, control the people.

The Chemistry of Comfort

All this talk of empires and microbes might make bread sound clinical, but let's be honest: bread is also comfort incarnate. There's a reason the smell of baking bread can sell houses and end arguments.

That smell? That's chemistry, too. When the crust turns golden brown, you're witnessing the Maillard reaction at work. Amino acids (*from proteins*) and reducing sugars (*from starch*) collide under heat, reorganizing into hundreds of aromatic compounds. Some smell nutty, others malty, sweet, or toasty. Together, they create the irresistible perfume of fresh bread, the olfactory equivalent of an embrace.

And the sound, yes, bread has a sound. Crack open a crusty loaf, and you hear the faint snap of starch polymers fracturing, releasing steam trapped in the crumb. Food scientists call it "acoustic crispness." You and I call it satisfying.

Bread as Controlled Fire

At its core, bread is about taming fire. Cooking grains transforms them, but baking bread pushes that transformation to the edge of combustion. Bakers walk a thin line between a golden crust and char. Too little heat, and the bread is dense and raw inside. Too much, and it's blackened carbon. The trick is a controlled burn: enough heat to gelatinize starch, coagulate proteins, and evaporate water without tipping into ash.

That control, fire calibrated to chemistry, is what makes bread not just food but science.

Bread in the Lab

In modern times, bread has entered the laboratory as much as the kitchen. Food scientists measure loaf volume using displacement tanks, examine crumb structure under a microscope, and map volatile aroma compounds using gas chromatography. Artisan bakers, too, have become scientists in their own right, tracking hydration percentages, fermentation times, and flour protein content with the precision of engineers.

Chad Robertson of Tartine Bakery, for instance, popularized naturally leavened bread by publishing fermentation schedules written like mission plans. Every baker with a sourdough starter is running a microbial ecosystem in a jar, feeding and culling populations like a biologist with Petri dishes.

And yet, despite the science, bread retains its mystique. No two loaves are identical. Ambient temperature, humidity, and even the microbial cloud of a particular kitchen leave fingerprints on the final loaf. Bread is reproducible, but never repeatable. That paradox keeps bakers humble.

The First Chemistry Set

If you want to understand why humans became chemists long before they became physicists or engineers, look no further than bread. In bread, you see the birth of controlled transformation: raw inputs reorganized into something new, governed by invisible rules yet accessible through trial and error.

Every time you slice a loaf, you're seeing archaeology and chemistry layered together. The scorch marks of Stone Age hearths, the fermentation vats of Egypt, the bakeries of Rome, and the sourdough counters of San Francisco are all part of the same experiment.

Bread is not just food. It's the story of us learning to bend molecules to our will. And if that's not science, nothing is.

Chapter Two – Fire in a Box

Let's start with the obvious: cooking begins with fire.

I mean, yes, you can "cook" food by fermenting it, salting it, or letting enzymes go to town. Still, nothing has rewired human biology and society quite like the act of pointing something edible at an open flame and waiting. Fire is the original lab instrument; both brutally simple and ridiculously hard to control.

Imagine you're a proto-human 300,000 years ago. You've just figured out how to keep a flame alive after lightning struck a tree. Congratulations, you've unlocked the cheat code for calories. Heat breaks down collagen in meat, making it more tender and easier to chew. It gelatinizes starch in roots, turning fiber into edible carbs. And it nukes pathogens, lowering the chance your dinner kills you before the saber-tooth tiger does.

That was step one: *fire, meet food*. Step two was all about containment.

Campfire Chemistry

Early humans didn't have Le Creuset or convection ovens with programmable fans. They had sticks, stones, and dirt. Cooking "equipment" looked like this: skewering hunks of meat on a stick and holding it near a flame; dropping tubers into the embers; or digging a pit, filling it with hot rocks, covering it, and slow-cooking whatever unlucky animal you caught.

Pit cooking, by the way, is still alive and well. Hawaiians call it an imu. New Englanders do a "clambake" the same way. It works because hot rocks release heat slowly and evenly, creating a primitive thermal battery. You bury the food, walk away, and come back hours later. Voilà, dinner, tender and smoky. It's a low-tech sous vide.

From Rocks to Pots

Fast-forward a bit. Around 20,000 years ago, humans in East Asia figured out pottery. That was a game-changer. A clay pot could hold water so that you could boil things in it. Boiling may not sound sexy, but chemically, it's a miracle. Suddenly, you're not just searing and roasting; you're extracting. Boiling coaxes nutrients from bones, dissolves starch into porridge, and unlocks flavors you can't get over a flame alone.

And let's not forget: boiling water kills microbes. That's not just food chemistry; it's survival chemistry.

Pottery also gave rise to stews, soups, and eventually the whole idea of "cuisine." Before pots, dinner was "whatever fits on a stick." After pots, dinner became "whatever flavors dissolve together in hot water."

Enter the Oven

Okay, so we've got fire, pits, and pots. Next up: ovens. Archaeologists in Central Europe have discovered clay ovens dating back 29,000 years; giant stone boxes where early humans roasted mammoth meat alongside tubers. But it was the Egyptians (*again!*) who perfected bread ovens. Their beehive-shaped clay ovens were heated by burning wood inside, then scraped clean for baking, setting the standard. You can still see their descendants across the Middle East today: the tandoor, the tabun, and the saj.

Why ovens? Because they let you trap heat. That's the difference between torching food and cooking it evenly. An oven is a controlled microclimate: hot air circulates around the food, heating it on all sides, while the walls store and radiate heat.

If campfires are chaotic, ovens are orderly.

Fire Evolves

Fast-forward again: medieval Europe. Castles, feasts, giant spits turning whole animals over open hearths. Kitchens were essentially indoor fire pits, as dangerous as they sound. Cooking was smoky, sooty, and hot. But clever cooks invented tools: wrought-iron grates, cauldrons hung from chains, and spits with crank handles (*sometimes powered by dogs running in treadwheels, yes, medieval kitchens had dog-powered rotisserie machines*).

This period gave us the fireplace hearth, which served as both a heater and a stove. In fact, the word "stove" originally meant only "heated room."

The Big Leap: Stoves

The real breakthrough came in the 18th century, when people grew tired of living in smoke. Enter the cast-iron stove. It was like moving from a bonfire to a reactor core. Suddenly, you had walls, grates, and vents to control airflow. You could add wood, close a door, and get consistent heat without burning your eyebrows off.

Ben Franklin even invented his own "Franklin Stove" to maximize heat efficiency (*though it was more for warming rooms than for cooking*). Still, the principle held: fire belongs in a box, not in your face.

By the 19th century, iron stoves had become standard in households. They featured multiple burners, ovens, and doors. You could boil, fry, roast, and bake simultaneously. Cooking shifted from brute-force survival to orchestrated multitasking.

Gas, Electricity, and the Great Kitchen Arms Race

We now turn to the modern era. In the mid-1800s, gas lines spread through cities, and gas stoves became the norm. They were faster, cleaner, and adjustable—no more hauling wood or coal. Just twist a knob and whoosh, flame on demand. Bakers loved gas for even heat, and cooks loved it for precise control.

Then electricity became available in the early 20th century. At first, electric stoves seemed like toys, with less heat and clunky controls. But they had one huge advantage: no open flame. Safe, predictable, and, in theory, more efficient. By the 1950s, electric stoves were marketed as futuristic marvels: push-button cooking for the space age.

Still, ask most chefs today, and they'll tell you: fire wins. Gas stoves give you instant, visual feedback; turn the dial, and the flame changes immediately. Electric coils and glass tops lag. Then came induction. Induction stoves don't heat the air or the coil; they use electromagnetic fields to heat the pan itself. It's fast, precise, and efficient. Induction is the Tesla of stoves: sleek, quiet, and a little smug.

Persistence of the Primitive

So here are, with every gadget at our disposal. Commercial kitchens bristle with combi ovens (*steam-and-convection hybrids*), salamanders (*overhead broilers hot enough to melt steel beams, or at least cheese*), and gas ranges built like tanks.

Home cooks can buy smart ovens with Wi-Fi, apps, and AI that tells you when your chicken is ready to serve.

And yet pit cooking survives. Wood-fired pizza ovens thrive. Cast-iron skillets never went out of style. The tandoor is as popular as ever. Why? Because these "primitive" methods do what modern gear can't.

Wood smoke imparts flavors no electric coil ever can. Clay ovens trap and radiate heat in ways stainless steel can't replicate. Cast iron holds heat like a thermal capacitor. Even the humble campfire, nothing beats it for steak char or marshmallow caramelization.

Cooking technology advances, yet the flavor remains stubbornly retro.

The Why of Persistence

Why do old methods stick around? Because cooking isn't just about efficiency. It's about flavor, ritual, and psychology. Gas stoves persist because chefs trust the flame. Wood ovens persist because they give bread that blistered crust. Campfires persist because we're hardwired to find comfort in glowing embers.

In other words, not every advance replaces the old. Sometimes it simply adds to the arsenal. We didn't abandon the wheel after inventing the airplane. We still use both—depending on the trip. The same applies to ovens and stoves: clay, iron, gas, electric, and induction. They all coexist, each optimized for different outcomes.

The Kitchen as Lab

Think of the modern kitchen as a multi-instrument laboratory. The oven is your controlled-environment chamber. The stovetop is your rapid-heating workstation. The smoker is your long-duration reactor. The microwave? A particle accelerator for water molecules. Each tool offers a distinct set of reactions to exploit.

That's why professional kitchens look less like cozy hearths and more like steel labs with flames. Temperature probes, timers, and blast chillers; this is chemistry scaled up and made repeatable.

The Future of Heat

Where does heating go next? Solid-state thermal technology, smart ovens that track chemical markers in real time. Imagine an oven that knows when Maillard reactions peak and automatically adjusts the heat. Or pans that measure pH shifts during caramelization and tell you when to flip your steak.

But don't worry, none of this will kill the campfire. Even if you invent the most advanced AI-driven cooking system, people will still gather around a flame, skewer something on a stick, and call it dinner.

Containment and Control

Cooking began in chaos: food over fire. It evolved into containment: pits, pots, and ovens. Then into control: stoves, gas, and electricity. Now it's precision: induction, smart sensors, and digital thermostats.

But through all those upgrades, the principle hasn't changed: cooking is applied chemistry through heat. The tools just give us better ways to manage the equation.

So next time you turn a knob on your range or slide a loaf into the oven, remember, you're part of a lineage that spans from Paleolithic embers to induction stovetops. You're not just cooking dinner. You're running the latest version of humanity's oldest experiment: fire in a box.

Chapter Three – The Flavor Hackers

If the previous chapter was all about heat, Chapter Three is about the inputs: the stuff we cook. Fire without food is just camping. Fire plus food is chemistry. But fire plus the right food is cuisine.

And humanity didn't stumble upon good ingredients in one afternoon. We hacked our way to them... by accident, by curiosity, by desperation. What we discovered along the way changed not only how things taste but also how long they last, how safe they are, and how societies organize around trade and conquest.

So buckle up. This chapter is part scavenger hunt, part chemistry lab, and part Indiana Jones movie, except instead of golden idols, we're chasing salt, pepper, saffron, and the choicest cuts of beef.

The Original Gangster

Let's start with the most essential ingredient in culinary history: salt.

Chemically, it's sodium chloride, NaCl. Structurally, it's a perfect little cube of ions. Biologically, it's essential that your neurons literally fire using sodium gradients. Without it, you die. Too much, and you die differently. The trick is the middle ground.

Early humans noticed that meat left near salty ocean spray didn't spoil as quickly. Eventually, we figured out that salt isn't just a flavor; it's also a preservative.

Salt pulls water out of cells (*osmosis*), making life miserable for bacteria. Without water, microbes can't multiply. Salted fish, salted pork, salted everything: suddenly, food could travel. Armies marched on salt. Empires rose because of salt. Roman soldiers were partly paid in it (*salarium, the root of the word "salary"*). Gandhi's march to the sea was about it.

But beyond politics, salt had another trick: it made food taste better. Salt enhances flavor by suppressing bitterness and amplifying sweetness and umami. A pinch of salt doesn't just make soup saltier; it makes it more itself.

Modern cooks think of salt as background noise. Wrong. Salt is the volume knob for your entire meal.

The First Global Obsession

Now, if salt is oxygen, basic, and necessary, pepper is nitrous oxide.

Black peppercorns are dried berries from a tropical vine native to India. To the ancient Romans, they were worth more than gold by weight. When Rome fell, pepper became the carrot dangling before European explorers. Vasco da Gama didn't sail around Africa for sunsets. He was chasing pepper. Columbus was seeking pepper when he accidentally found the Bahamas.

Why pepper? It doesn't preserve food like salt, but it adds heat, aroma, and complexity. The active compound is piperine, a pungent alkaloid that activates the same receptors as chili peppers, though less intensely. Sprinkle it on bland meat, and suddenly you've got depth.

Pepper rewired global trade routes. It made Venice wealthy, Portugal ambitious, and England imperial. All for tiny black beads that, chemically speaking, do one simple thing: irritate your mouth in a way you enjoy.

Herbs and Spices: The Chemical Arsenal

Once humans realized plants could season food, we went wild for spices like cinnamon, cardamom, cloves, and nutmeg. Every spice is a chemical weapon that plants have evolved to repel insects. Humans came along, tasted those defenses, and said, "Delicious. Let's build an economy around this."

Take cinnamon, harvested from tree bark and rich in cinnamaldehyde, which delivers both flavor and antibacterial properties. Or chili peppers: capsaicin evolved to discourage mammals from eating them. Humans not only ignored the warning but also bred peppers to make them hotter. The Scoville scale is basically us bragging about how much pain we can tolerate in pursuit of pleasure.

Spices did two critical things for early cooks:

They masked the off-flavors of preservation. Salted fish still tastes like salted fish. Add cloves, and suddenly it tastes like dinner, not   like   survival.

They have diverse cuisines. Mediterranean food tastes different from Indian food mainly because of the spice palettes. That wasn't geography; it was chemistry, packaged in plant oils.

Okay, Let's Talk Fat

Fat is flavor's bestie. OMG, totally.

Fat is caloric gold; nine calories per gram, more than twice the calories of carbs or protein. For early humans, fat meant survival through winter. For modern cooks, fat means flavor transport. Aromatic compounds dissolve more readily in fat than in water, so butter doesn't just taste good; it's a solvent for flavor.

Think of garlic sautéed in oil. The allicin compounds in garlic bind to fat molecules, hitching a ride to your taste buds in a way raw garlic can't. Or consider the marbling in steak: those thin intramuscular fat veins melt during cooking, basting the meat from the inside and delivering both juiciness and flavor.

Different fats bring different chemistry. Olive oil contains monounsaturated fats that withstand heat. Butter has milk solids that brown, developing nutty flavors. Lard gives pastries flakiness. Duck fat turns potatoes into a drug.

Fat isn't just fuel; it's the medium in which flavor lives.

The Anatomy of Taste

Now we get to the fun part: the discovery that not all meat is created equal.

A cow isn't just "beef." It's a biochemical map of muscles, each with its own texture, fat content, and connective tissue. The tenderloin? Barely used, so its muscle fibers are delicate and it melts in your mouth. The brisket? Tough, full of collagen. But simmer it long enough, and the collagen breaks down into gelatin, turning shoe leather into heaven.

Humans figured this out the hard way: trial and error over millennia. Chew a raw flank steak, and you'll crack a molar. Simmer it low and slow, and you'll get a stew that sticks to your ribs. That's chemistry: collagen $\rightarrow$ gelatin at $\sim160°F$, a transformation that gave birth to barbecue, braises, and comfort food everywhere.

Seafood works the same way. Fatty fish like salmon contain omega-3s that amplify umami. Shellfish are rich in amino acids, including glycine, which activate sweetness receptors. Ancient coastal communities learned quickly: boil clams with herbs, roast fish on hot stones, and ferment shrimp into pastes. Each method exploited chemical differences among species and cuts.

The "discovery" wasn't just about which animals to eat, but how to cook each part for maximum payoff.

The Vegetable Revolution

We can't leave plants out. Humans learned early that not all roots, leaves, or seeds are equal. Wild brassica (*a scraggly weed*) gave us kale, cabbage, broccoli, and Brussels sprouts, all bred by farmers who saw potential in bitterness and texture.

Tomatoes, native to the Americas, were initially feared in Europe (*nightshade family = poison!*) before becoming a cornerstone of Mediterranean cuisine. Likewise, potatoes went from Andean staple to European salvation during famines.

Cooking plants does two big things: breaks down cell walls and unlocks nutrients. Heat ruptures cellulose, releasing sugars and softening texture. It also makes nutrients like lycopene in tomatoes and beta-carotene in carrots more bioavailable.

So when humans boiled, roasted, and fried their veggies, they weren't just making them tastier. They were improving nutrient absorption, inadvertently hacking biochemistry.

Controlled Rot

No chapter on "discovery" is complete without fermentation. At some point, someone left milk out too long, drank the funky liquid, and, miracle of miracles, it didn't kill them. Yogurt was born. The same goes for grapes (*wine*), barley (*beer*), and cabbage (*sauerkraut*).

Fermentation was humanity's first microbial partnership. Yeasts and bacteria broke down sugars and proteins, producing acids, alcohols, and gases that preserved food and transformed flavor. Soy sauce, kimchi, miso, and cheese are all accidents we leaned into.

It's not just preservation; it's enhancement. Aged cheese isn't just "old milk." Proteins break down into peptides and amino acids, fats oxidize into complex aromas, and bacteria generate flavor molecules you can't produce with heat alone. Parmesan contains butyric acid, isovaleric acid, and glutamates; it's an orchestra of chemistry.

Fermentation proved that the best way to cook is sometimes to wait.

Modern Flavor Hacking

Today, we're spoiled. We can walk into a Kroger and choose from hundreds of spices, cuts, and oils our ancestors would've considered black magic. But the same principles apply: chemistry plus curiosity equals cuisine.

Molecular gastronomy is just the latest chapter. Ferran Adrià at El Bulli foamed carrots, froze olive oil into a powder, and made diners rethink texture. Heston Blumenthal found that adding anchovies to beef stew turbocharged the umami. Samin Nosrat boiled it down to four elements: salt, fat, acid, and heat.

We're still discovering. Koji mold, used for centuries in Japan, is now appearing in charcuterie and even vegan cheese. Seaweed is being harvested for its umami-enhancing compounds. Even insects are on the horizon, packed with protein and chitin, waiting for the right chef to turn them into craveable snacks.

Ingredients Matter

Here's the big takeaway: cooking isn't just about fire. It's about what you choose to put in the fire. Salt turned survival food into cuisine. Spices built trade routes. Meat cuts gave rise to barbecue. Vegetables and fermentation enhance flavor and nutritional value.

Every discovery shifted the balance among chemistry, culture, and survival. And every time we cracked a new ingredient, we weren't just feeding ourselves—we were evolving our relationship with food.

Conclusion: The Hacker's Legacy

Think of early cooks as hackers. They poked, prodded, and tested. They salted meat, steeped leaves, boiled bones, and gnawed on roots. Ninety percent of the time, the results were disgusting. But the ten percent that worked? Those discoveries became the backbone of global cuisine.

The pantry is a museum of successful experiments. Every jar of cinnamon, every cut of steak, every wedge of cheese is proof that someone, somewhere, played with chemistry until flavor clicked.

And we're still hacking.

The next time you sprinkle salt, grind pepper, drizzle oil, or slow-cook brisket, remember: you're standing on the shoulders of millennia of curious, hungry scientists without lab coats, just fire, instinct, and the stubborn belief that food could always taste better.

Chapter Four – Rice and Rye: Twin Experiments

Here's the thing about grains: they look boring. Seeds. Little pebbles you sweep off the floor. And yet, two of them, rice and rye, wrote human history in very different fonts. Rice fed empires. Rye fueled survival where rice wouldn't dare grow. Together, they prove that ingredients aren't just food; they're civilization's operating systems.

The Civil Engineer of Grains

Let's start with rice. If wheat is the loud, extroverted sibling of the grain family (*bread, pasta, pizza!*), rice is the quiet, studious one. It doesn't look like much, but it feeds more than half the planet.

Rice domestication began around 9,000 years ago in the Yangtze River basin. Picture this: ancient farmers wading through swampy paddies, ankle-deep in muck, slowly realizing that flooding a field kills weeds while rice survives—boom, the first example of humans weaponizing ecology.

Rice didn't just spread across Asia; it pulled off a hostile takeover, armed with a few suspiciously good selling points. Its primary "killer features" included insane caloric efficiency, because who doesn't love cramming as many people as possible onto one hectare of land? And the "built-in irrigation system" was pure genius, doubling as a luxurious, all-inclusive resort for fish... and, regrettably, the occasional malaria-carrying mosquito, a real two-for-one deal for ecosystems.

Chemically speaking, the rice grain is fundamentally starch, with starch, and, of course, more starch thrown in for good measure. Around 80% of the grain is merely glucose molecules impatiently waiting for a dip in hot water.

But as any proper cook knows, not all starches are created equal, and the type of starch makes all the difference. The grain contains two varieties: the straight-laced amylose (*long, unbranching chains that are too proper to gel easily*) and the more boisterous amylopectin (*branchy, sociable chains that readily turn into a delightful, sticky mess when heated*). The precise ratio between these two determines whether you'll end up with a proper, fluffy affair à la basmati or a rather clingy situation, perfect for sushi.

Ancient cooks figured this out without molecular diagrams. Want separate grains? Go long-grain, high-amylose. Want sticky rice balls? Go short-grain, high-amylopectin. Science, meet dinner.

Rice in Action

Ah, the humble grain, Oryza sativa, a seemingly innocuous staple that is, in fact, a high-stakes, high-wire act of kitchen chemistry. It's an arena where culinary triumph can bemeasured in perfectly fluffed grains, and utter failure in something suspiciously resembling thick gruel.

Behold, the "chemistry in action." You think you're just adding water and applying heat, a simple equation? Nonsense. You're initiating a ballet of physics, a delicate dance of molecular forces and thermal dynamics. There are a lot of variables. Absorption: the silent overture. Water doesn't just "seep"; it infiltrates the starch granules with the stealth of a spy, preparing for the main event.

Another is Gelatinization: the dramatic climax. Heat, the great hydrogen-bond breaker, forces the granules to swell like over-inflated egos at a Manhattan cocktail party. They burst open, softening into the tender texture we so desperately seek.

Lastly, Steam Puff: The grand finale. At the boiling point, pressure builds like a diva's tantrum, fluffing the grains into airy perfection. Love a good steam!

The moral of this microscopic drama? Nail the ratios, and you are a culinary god or goddess. Get them wrong, even by a smidge, and your dinner is doomed to be a sticky, gloppy tragedy worthy of a Greek playwright. Good luck, and may your gelatinization be ever successful.

The best part? Rice is the gateway to culinary chemistry. Add fermentation, and you've got sake. Add high-heat oil, and you've got fried rice, where starch molecules undergo Maillard reactions, turning yesterday's leftovers into dinner with flavor steroids.

Rice as Civilization's Backbone

Rice didn't just feed people; it organized societies. Irrigation required massive coordination. You can't flood paddies without dikes, canals, and labor schedules. That meant centralized governments, tax records, and armies to protect the fields. In short: no rice, no dynasties.

Even today, rice is more than just food; it's a culture. Japanese Shinto rituals use rice wine, and Indian festivals revolve around rice dishes. In much of Asia, the word for "meal" literally means "rice." Forget bread; this is the staff of life.

The Underdog Grain

Now let's talk rye. If rice is the smooth city planner, rye is the scrappy frontier survivalist. It thrives where wheat sulks, in cold climates, poor soils, and at high altitudes. Consider Northern Europe, Siberia, and other regions where farming feels like a punishment.

Rye didn't start as anyone's first choice. Archaeological evidence suggests it began as a weed in wheat and barley fields. But when the climate cooled, wheat failed, and rye shrugged and said,

"No worries, I got this." Farmers noticed. By the Iron Age, rye had carved out its niche as the grain of last resort and eventually became a staple.

Chemically, rye is a rebel. Its starch is softer than wheat's, making it less suited to airy loaves. Its gluten is weaker, so you don't get the same elastic network to trap CO_2. Result? Dense bread. Heavy, earthy, sometimes sour.

But rye compensates with pentosans, complex sugars that absorb water like sponges. That's why rye bread stays moist for days, unlike wheat bread, which goes stale faster than an open bag of chips. With natural fermentation (*sourdough*), rye becomes hearty, tangy loaves that powered peasants through Russian winters.

Rye's Dark Side

Of course, rye has baggage. Enter the ergot fungus. This little parasite thrives on rye and produces alkaloids that, in small doses, cause hallucinations and, in large doses, death. Medieval outbreaks of "St. Anthony's Fire" (*ergot poisoning*) caused convulsions, gangrene, and visions. Some historians even suggest the Salem witch trials might have been fueled by ergot-contaminated rye bread.

So rye gave with one hand (*calories, resilience*) and took with the other (*hallucinations, necrosis*). Humanity, being humanity, kept eating it anyway, because the alternative was starvation.

Rye as Flavor Identity

Rye bread doesn't get the global spotlight as rice does, but it builds identity. Eastern European cuisines revolve around it. Think pumpernickel in Germany, black bread in Russia, and caraway-studded loaves in Poland. Rye gave these regions their culinary DNA; dense, sour, resilient food for dense, sour, resilient climates. Perfect for people who are sometimes dense, sour, and resilient.

And here's the kicker: rye's distinct flavor comes from ferulic acid breaking down during fermentation, which releases compounds like 4-vinyl guaiacol. That's the smoky, spicy note you taste in rye whiskey, too. So rye not only gave us bread but also booze with attitude.

Rice vs. Rye: A Tale of Two Chemistries

Two grains can teach us a masterclass in strategy, really. Observe, if you will, a side-by-side comparison of two botanical philosophies.

Rice is a sun-worshipping sybarite, thriving in perpetual spa-like conditions (*warm, wet*). Starch is its game; fluffy versatility is its calling card. It's an empire-builder, a social climber that constructs palaces and paperwork. It's optimized for abundance, and frankly, we're still cleaning up after its bureaucratic impulses.

Rye, the other side of that coin, is the stoic, slightly morose Northern cousin. It prefers a bracing chill and poor soil. Lacking the social graces of strong gluten, it makes do with water-absorbing pentosans, the botanical equivalent of pulling up its collar against the weather. It is dense, sour, and stubbornly survival-driven, the quiet backbone of villages enduring endless winters.

Chemically speaking? Rice screams: "Maximize digestible starch! We feast!" while Rye mutters: "Maximize resilience. We persist."

Both, in their own way, managed to turn raw calories into the foundations of human culture. One built the feast hall; the other just made sure you didn't starve in the dark.

Cooking the Twins

Cooking with rice and rye today is like inheriting two very different toolkits.

Rice lets you experiment with texture: steamed, fried, fermented, or puffed. Want chemistry fun? Try puffed rice: you heat the grains under pressure, then release them. Steam explodes outward, and the starch sets into a crunchy foam. That's a cereal aisle in one reaction.

Rye is about depth: long fermentation, dense loaves, and earthy flavors. A rye starter culture is a microbial zoo, with Lactobacillus species dominating and producing a lactic-acid tang that wheat breads rarely achieve.

Together, they remind us that cooking isn't just about "can we eat this?" It's about "how do we make this edible *and* enjoyable in the environment we live in?"

Conclusion: The Grain Equation

Rice and rye aren't just ingredients. They're solutions to different constraints. Rice solved for efficiency and abundance in warm, wetlands. Rye solved for survival and stamina in cold fields. Both shaped entire civilizations by dictating what was possible on a plate and, by extension, in a society.

So next time you scoop fluffy rice or bite into a slab of dark rye, remember: you're not just eating grains. You're eating the solutions to ancient chemical problems, written in starch and protein, tested in fire and water, and debugged by millennia of hungry humans.

And if that doesn't make your dinner feel more important, nothing else will.

Chapter Five – The Science of Implements

Heat in the kitchen has led to many time-saving innovations, but one has received very little attention: the humble oven mitt.

Humanity's fabric-based defense against molten death, a domestic hero born of necessity and questionable fashion sense

Its story begins not with a tailored glove but likely with a folded rag, perhaps a piece of repurposed animal hide, anything to avoid touching the scorching Roman kiln or medieval hearth. For centuries, this crude hand protection remained largely unchanged, a testament to our ancestors' tolerance for third-degree burns and general lack of imagination.

The breakthrough came with industrial textile production, leading to the quilted, insulated "glove" we barely notice today. By the 1950s, the mitt had fully embraced domestic life, evolving from purely utilitarian protection into a canvas for questionable taste: adorned with sunflowers, witty sayings, or a cat's face.

The introduction of silicone in the late 20th century marked a material revolution, offering greater heat resistance but often at the expense of cloth's ergonomic flexibility.

From a simple linen scrap to a heat-proof silicone puppet, the oven mitt endures as a silent, often singed, testament to our desire for perfectly baked cookies and unblistered fingers. A truly gripping tale.

Fire may be the heart of cooking, but implements are its hands. Every knife, pot, and spoon is a tool designed to solve a scientific challenge: how to cut more cleanly, transfer heat more evenly, or deliver food more safely to the mouth. Trace their evolution, and you see not only cultural shifts but also the gradual discovery of principles in physics, chemistry, and biology, often centuries before anyone wrote down the equations.

The Physics of Sharpness

The earliest cooking tool was not the pot or the plate; it was the blade. Flint chips, millions of years old, show that our ancestors grasped the concept of sharpness long before metallurgy. A sharp edge is simply matter concentrated into a narrow point, reducing the area of contact. Pressure equals force divided by area, so when the area shrinks, pressure spikes. That's why a chipped stone can easily part an animal's hide.

Metals took this principle further. Bronze, then iron, and eventually steel enabled blades to be thinner, stronger, and more durable than stone blades. Heat treatment altered the microstructure of iron alloys, shifting the balance between martensite and pearlite and producing edges that held stress longer. Even without metallurgical theory, smiths empirically tuned carbon content and quenching methods, turning metallurgy into applied kitchen science.

Cutting isn't just mechanical; it's chemical in its consequences. A sharp knife cleanly severs plant cells, slowing oxidation and enzymatic browning. Dull blades crush and rupture cells, flooding tissues with enzymes that accelerate discoloration and bitterness. That's why onions sting more with a blunt edge and why apples brown faster. Every slice is biochemistry in action.

Geometry Meets Thermodynamics

Once the pots arrived, scooping became an engineering problem. Shells and hollowed bones gave way to carved spoons and, eventually, ladles. Their design solved multiple physical problems:

the geometry of a concave bowl holds liquid by balancing gravity and surface tension; the long handle keeps human skin outside the danger zone of boiling water.

The whisk, which emerged in early modern Europe, exemplifies the application of fluid dynamics. Multiple thin wires introduce shear forces into liquids, uncoiling proteins in egg whites and cream. Those proteins denature and realign around trapped air bubbles, forming foams. Whipping cream is less about "adding air" than about restructuring proteins into scaffolds. The whisk is, in essence, a protein-folding machine.

Materials as Thermal Solutions

Clay pots were the first controlled cooking chambers. Clay is a poor conductor but an excellent insulator: it distributes heat slowly, reducing hotspots and providing a steady simmer. As metallurgy advanced, bronze and copper changed the rules.

Copper, my friend, is basically the Usain Bolt of the metal world when it comes to moving heat. While other materials are strolling along, taking their sweet time, copper is sprinting, a speed demon for warmth. This means heat zips through it incredibly quickly, almost as if it's racing against time itself. Food scientists today still rank it the gold standard for avoiding hotspots.

But copper's reactivity with acidic foods leaches metallic flavors, which is why a tin lining or, later, stainless steel is used. The engineering solution? Composite materials, layering conductive metals with inert ones. Modern cookware is literally a sandwich of metallurgical science.

Cast iron is the father of an old friend who has never taken a day off from work. That means cast iron has a high thermal mass. It's like comparing a cozy campfire (*your thin pan*) to the sun (*your cast iron*); when we talk about the heat they hold, there's no contest.

Cast iron stores an enormous amount of heat. That makes it perfect for searing. Once hot, it stays hot, even when a cold steak hits the surface. Cast iron is less about conductivity than about heat storage. It is the kitchen's culinary battery.

Biology and Ergonomics

The fork and chopsticks may seem cultural, but their designs are rooted in biomechanics. Chopsticks extend the human hand's pincer grip, exploiting leverage for delicate handling. Their length also solves a thermodynamic issue: distance from heat. A pair of sticks turns burning-hot stir-fry into finger-safe portions.

Forks, by contrast, amplify puncture force. The tines concentrate pressure at points, the same physics as sharp blades. A fork is essentially a miniature spear, optimized for portioning and stabilization. Both tools reflect applied ergonomics, the study of how human anatomy meets physics

The Psychology of Surfaces

Plates began as mere surfaces of wood, clay, or metal, but their scientific significance lies in perception. Porcelain, perfected in Tang dynasty China, is chemically nonporous, resisting absorption and contamination. But its whiteness is equally important: it reflects light across the entire visible spectrum, making food colors appear more vibrant and saturated.

Modern sensory science confirms what ancient hosts intuited: plate color and shape alter flavor perception. A study published in the Flavour journal found that desserts plated on round, white dishes were rated 10% sweeter and more pleasant than identical desserts on black, square plates. This isn't mysticism; it's cognitive psychology. The plate is part of the flavor system.

From Ritual to Precision

The rise of measuring spoons and cups marks the entry of quantification into the kitchen. Originally, "a spoon" meant whatever utensil was at hand. Over time, standardization (*a teaspoon = 5 mL, a tablespoon = 15 mL*) transformed recipes from oral tradition into a reproducible protocol. This was kitchen science aligning with laboratory practice.

Thermometers took it further. In the 19th century, confectioners realized that boiling sugar required precise temperature stages, such as thread, soft ball, and hard crack, each a window into the chemistry of sucrose polymerization and caramelization. Today, digital probes give cooks the same accuracy as lab scientists, measuring protein denaturation in meat or precise extraction temperatures in coffee. Implements became instruments.

Small Tools, Big Science

Peelers embody leverage. A blade set at controlled angles removes thin layers of skin with minimal waste, using geometry to maximize efficiency. Tongs exploit mechanical advantage, converting hand force into clamping force at a distance. Even the humble grater is microengineering: hundreds of sharpened perforations that dramatically increase surface area, altering reaction kinetics in cooking (*grated cheese melts faster, and grated ginger releases more aromatic oils*).

These tools are not trivial. They are incremental optimizations that accelerate, improve cleanliness, and increase consistency in chemical and physical transformations.

Implements as Scientific Evolution

Examine the kitchen from a scientific perspective, and each tool becomes a tangible hypothesis. A flint knife serves as a hypothesis: its sharp edges amplify pressure. A clay pot serves as a hypothesis, too, as its insulating properties create stable heat. A copper pan represents a hypothesis in its ability to conduct heat, reducing hotspots. A whisk can be viewed as a hypothesis, in which shear forces help restructure proteins.

Forks and chopsticks serve as hypotheses; they extend our anatomy, allowing us to better control temperature and portion size. A porcelain plate represents a hypothesis on how its surface alters our perception of food. Lastly, a thermometer acts as a hypothesis, with its precision transforming guesswork into reproducible measurements.

Each innovation was an experiment, tested over generations and replicated until it became a tradition.

The Lab in the Kitchen

Implements are not accessories to cooking; they are its scientific infrastructure. They embody metallurgy, geometry, thermodynamics, biology, and psychology, translated into wood, clay, and steel.

When you slice with a chef's knife, whisk egg whites, or sear in a cast-iron pan, you are conducting experiments with tools refined over centuries of trial and error. They are solutions to problems you didn't even know you had: how to apply pressure, manage heat, manipulate protein, or even trick your brain into tasting sweetness.

The kitchen is a lab, the cook is a scientist, and every implement is a piece of accumulated knowledge made tangible. In this sense, cooking is the longest-running scientific experiment in human history, still ongoing, still evolving, one tool at a time.

Chapter Six – The World's Favorite Chemical

Okay, let's get something straight: sugar is a molecule. That's it. $C_{12}H_{22}O_{11}$. Two monosaccharides (*glucose and fructose*) bonded together in a neat little disaccharide handshake. Break the bond, and you release energy. Eat enough of it, and your brain throws a confetti parade.

So how did this one molecule become the planet's most beloved, most vilified, and most economically explosive ingredient?

Strap in. Sweetness isn't just a taste; it's chemistry, biology, history, and maybe even destiny.

The Chemistry of Sweet

Why do we even like sweets? Blame evolution. Humans are wired with receptors on the tongue (*T1R2 and T1R3 proteins*) that detect sugar molecules. When sugar docks, those receptors send fireworks to the brain's reward system. Translation: this is energy, so eat it.

Because glucose equals life. It's the body's preferred fuel. Your cells run on it, and your brain gulps it like a sports drink. Early humans who craved sugar survived better. They chased down ripe fruit and honey, not bitter, possibly poisonous leaves. Sweetness is survival disguised as pleasure.

Fructose, the other half of table sugar, is even sweeter per gram than glucose. Combine them, and you've got sucrose, stable enough to crystallize, portable enough to ship, and tasty enough to addict entire civilizations.

Plants with a Secret

Sugarcane was domesticated in New Guinea around 8,000 BCE and spread through India and Southeast Asia, eventually reaching the Middle East. Ancient Sanskrit texts describe boiling cane juice into crystalline chunks called khanda, yes, the root of the word "candy."

Sugar beets emerged much later, in 18th-century Europe, but they showed that sugar wasn't exotic; it was everywhere, hidden in plants, waiting to be extracted by chemistry. Slice, boil, evaporate, boom, and crystals.

Extraction is science in action: heat drives off water, sucrose molecules lock together into lattices, and suddenly you have something entirely new: solid sweetness.

Sugar in the Your Kitchen Lab

Cooking sugar turns it into a chemistry demo gone rogue. Heat sucrose past 320°F, and you hit caramelization; molecules break, recombine, and spin off hundreds of new compounds:

diacetyl (*buttery*), maltol (*toasty*), furans (*nutty*). That brown, glossy caramel sauce is literally sugar shattered and reassembled.

Add proteins to the mix, say, in crème brûlée, and the Maillard reaction joins the party. Sugars react with amino acids, creating complex aromas that lean savory. That's why the crackly top of crème brûlée tastes deeper than plain caramel. It's sugar doing molecular improv with protein.

Sugar is also functional: it lowers water activity in jams, starving microbes. It stabilizes foams (*meringue*), tenderizes gluten (*cakes*), and adds bulk to ice cream, lowering the freezing point so it stays scoopable. Sugar isn't just sweet; it's the Swiss Army knife of kitchen chemistry.

Sugar's First Conquests

Now, the history. In India, sugar was medicine before it was dessert. Arab traders carried it west, refining the extraction into an art. By the Middle Ages, "sweet" was a rarity in Europe, a luxury reserved for kings and popes. Honey was nice, but sugar was refined, crystalline, and eternal.

Then came colonialism. Europeans realized sugarcane thrived in tropical climates. The Caribbean, Brazil, and the American South became sugar-producing regions. Plantations, enslaved people, and ships turned sugar into the engine of the Atlantic economy. Coffee, tea, and chocolate, all bitter on their own, became palatable only with sugar. Suddenly, daily life revolved around it.

The British Empire ran on three things: coal, steel, and sugar. Afternoon tea wasn't a dainty ritual; it was a caffeine-sucrose productivity hack. Industrial workers gulped sweet tea for cheap calories, fueling factories. Sugar wasn't just a taste; it was an industrial energy policy.

The World has a Sweet Tooth

By the 19th century, sugar was everywhere. Prices fell. Beets democratized supply. Sweetness shifted from royal luxury to a working-class staple. The human palate, shaped by scarcity, suddenly drowned in abundance.

Cue the rise of dessert as a category. Cakes, pastries, and puddings, whole industries, bloomed. Chocolate bars appeared, blending cocoa's bitterness with sugar's charm. Soda fountains bubbled with carbonated sugar water. If bread was the backbone of civilization, sugar was becoming its nervous system; fast, addictive, omnipresent.

The Biology of Addiction

Let's talk about why you can't stop after "just one cookie." Sugar triggers dopamine release in the brain's reward circuitry, the same system activated by drugs such as nicotine and cocaine. Each bite reinforces craving. Unlike fat or protein, sugar acts quickly. Glucose surges into the bloodstream, insulin spikes, energy floods in, and the brain records: do this again.

The problem? Evolution wired us for scarcity. Finding fruit once a week was a great survival strategy. Eating donuts daily leads to metabolic chaos. Obesity, diabetes, fatty liver, diseases of abundance. The biology that once saved us now betrays us.

Sugar as Villain

By the late 20th century, sugar had become public enemy number one. Nutritionists blamed it for obesity epidemics, heart disease, and even cancer. Governments launched campaigns to reduce added sugar. Food scientists scrambled to invent substitutes: aspartame, sucralose, and stevia. Each promised sweetness without calories, sparking controversy.

And yet, the demand never stopped. Global sugar consumption keeps rising. Because, for all our warnings and substitutes, the taste of sugar is still the most primal green light our brains know.

Sweet Innovations

Modern kitchens don't just sprinkle sugar; they treat it like a lab reagent. Pastry chefs work with isomalt, a sugar alcohol that resists crystallization, to sculpt transparent showpieces. Ice cream makers balance sucrose, glucose, and corn syrup to fine-tune texture and freezing curves. Molecular gastronomists spin sugar into foams, spheres, and powders.

Even the soda industry is chemistry in action: precise ratios of high-fructose corn syrup, phosphoric acid, and carbonation create that addictive fizz. The can is less a beverage than a flavor equation, engineered for maximum craving.

Sugar in Science

Here's the kicker: sugar isn't just a culinary star. It's central to biology. DNA itself has a sugar backbone (*deoxyribose*). ATP, the energy currency of cells, is adenosine triphosphate, built on sugar. Carbohydrates aren't indulgences; they're the scaffolding of life.

In the lab, sugar solutions feed microbes during fermentation, support cell growth in culture, and stabilize proteins during storage. Bakers may see sugar as a source of sweetness, but biologists see it as structural fuel. Same molecule, two worlds.

The Future of Sweet

Where do we go from here? Scientists are exploring new frontiers: "rare sugars" like allulose, which taste sweet yet are metabolized to a lesser extent. Designer proteins like miraculin, from miracle berries, literally rewire taste buds to perceive sour as sweet. Imagine eating a lemon that tastes like candy. That's not sci-fi. It already exists.

Meanwhile, biotech companies engineer yeast to produce stevia compounds without cultivating plants. Sweetness is becoming programmable, stripped of its agricultural origins and rebuilt molecule by molecule.

But one truth remains: we'll never stop craving it. Our brains are locked in. The future isn't about escaping sugar; it's about outsmarting it.

The Empire of Sweet

Sugar isn't just an ingredient. It's a molecule that hijacked our biology, reshaped our economies, and shaped our cultures. From Indian cane fields to Caribbean plantations, from French patisseries to American soda cans, it has been both a delight and a disaster.

Every spoonful is chemistry; crystals dissolving, bonds breaking, receptors firing. Every craving is evolution whispering: calories, now.

So the next time you stir sugar into your coffee or bite into caramel, remember: you're not just indulging. You're participating in one of humanity's longest, strangest, and sweetest experiments.

Chapter Seven – Fermentation: Civilization's Pet Microbes

Here's the truth: fermentation is just rot we enjoy.

That's it. Microbes eat sugars, excrete byproducts, and we say, "Hmm, this tastes good, let's keep doing it." Somewhere between leaving grapes in a bowl and opening a frothy beer, humans realized that not all spoilage is bad. Some of it is delicious, and more importantly, some is safe. Fermentation is humanity's first biotechnology, long before we knew what "bio" or "technology" meant.

Nature's Happy Accident

Fermentation predates humans. Wild yeasts, single-celled fungi such as Saccharomyces cerevisiae, float everywhere: on fruit skins, in the air, even on our hands. When they land on something sugary, they do their thing: break down glucose and fructose, producing ethanol and carbon dioxide. Chemistry time: $C_6H_{12}O_6 \rightarrow$

2   C_2H_5OH +   2 CO_2

One molecule of sugar goes in, and two molecules of ethanol (*alcohol*) and two of CO_2 come out. It's neat, clean, and incredibly useful.

Animals have been noticed, too. Birds and monkeys have been observed getting tipsy on fermented fruit. But humans, predictably, decided to take it to the next level.

***

From Accident to Intention

The earliest archaeological evidence of intentional fermentation dates back about 9,000 years in China, to pottery shards bearing traces of a mixture of rice, honey, and fruit wine. Around the same time, barley beer appeared in the Middle East.

Different continents, same idea: let nature run wild on sugars, then drink the bubbly result.

But here's the crucial leap: once people realized fermentation was repeatable, they began treating it as controlled chaos. Clay jars became dedicated brewing vessels. Vats were reused, so resident yeast populations dominated. Civilization didn't just discover booze; it discovered microbial farming.

Wild vs. Cultured

This is where "naturally occurring" diverges from "man-made science."

Wild fermentation = rolling the dice. You leave grapes out, and yeast shows up. Sometimes you get wine; sometimes vinegar; sometimes sludge that smells like feet.

Cultured fermentation is like stacking the deck. You save dregs from a successful batch and pitch them into the next one. Eventually, humans began isolating and cultivating specific yeast strains. Modern brewing is essentially domestication: just as we have tamed wolves into Labradors, we have domesticated wild microbes into predictable, beer-making pets.

Natural fermentation has given us diversity; funky lambics, sourdoughs, and kimchi with distinct personalities. Controlled fermentation has given us consistency; lagers, IPAs, and wine that doesn't blind you. Both matter. One is art through chaos; the other is science through design.

Liquid Bread

Let's start with beer because it's essentially edible chemistry in a pint glass.

Step one: malt grains. That means soaking barley until it germinates, then drying it. Germination activates enzymes, amylases, that break down starch into simpler sugars. Beer isn't just about yeast; it's about preparing sugars in a form yeast can digest.

Step two: mash. You soak malted grain in hot water, letting enzymes break down starches into maltose. Think of it as pre-chewing for yeast.

Step three: boil with hops. Hops add bitterness (*thanks to alpha acids*) and antibacterial properties. They're like chemical bouncers, keeping spoilage microbes at bay.

Step four: fermentation. Yeast converts maltose into ethanol and CO_2. That's the fizz in your beer and the buzz in your brain.

Beer is a symphony of enzymology, microbiology, and thermodynamics, discovered by accident and refined into an industry.

Yeast Meets Grapes

For most of 2024, I lived in Walla Walla, Washington. While there, I visited at least one vineyard, tasting room, or winery every day. Walking through vineyards with the winemakers and volunteering during harvest and crush, I became enamored with the art and science of winemaking.

Wine skips starch conversion. Grapes already provide glucose and fructose, like a yeast buffet. Crush the grapes, let them sit, and the yeast goes to work. CO_2 bubbles, ethanol rises, and pigments from grape skins dissolve into the liquid.

Here's the neat part: grape skins also contain tannins, polyphenolic compounds that act as natural preservatives. So fermentation, together with tannins, made wine stable enough to age. Suddenly, wine wasn't just buzz juice; it was a commodity, tradeable across distances and decades.

Wild fermentation dominated early wines, but by the late 19th century, Louis Pasteur showed that yeast wasn't magic; it was biology. That unlocked modern oenology: selecting strains for predictable fermentation, controlling temperature to modulate flavor, and killing rogue microbes with sulfites. Wine evolved from a wild meadow experiment into a controlled biochemical process.

Fermentation You Can Chew

Alcohol gets the spotlight, but bread is fermentation's quieter triumph. Sourdough is a microbial team of yeast and lactic acid bacteria. Yeast produces CO_2, which puffs up the dough. Bacteria produce lactic and acetic acids, adding tang.

Gluten's elastic network traps CO_2 bubbles until the heat of baking locks them in place. It's the same reaction as in beer, except instead of fizz in a glass, you get air pockets in a loaf.

Bread showcases the dual face of fermentation: practical (*aerated structure*) and sensory (*complex flavor*). Without fermentation, flour plus water equals paste. With fermentation, flour plus water equals civilization's favorite food.

Beyond Booze and Bread

Alcohol and bread hog the glory, but fermentation touches almost everything: cheese, milk, lactic acid bacteria, and rennet enzymes, which yield curds and whey. Aging introduces molds and secondary microbes that generate flavors like nutty, funky, or sharp.

Pickles. Cucumbers soaked in brine invite lactic acid bacteria to colonize, producing acids that preserve and sour the cucumbers.

Soy sauce. A microbial marathon: soybeans inoculated

with Aspergillus oryzae mold, then brined for months while bacteria and yeast generate glutamates (*umami bombs*).

Kimchi. Cabbage + salt + time. Lactic acid bacteria break down sugars, creating tang, fizz, and microbial diversity that shifts week to week.

Every culture has fermentation at its core because preservation was essential long before refrigeration. Salt and microbes turned perishables into food that lasted months or years.

The Man-Made Era

For most of history, fermentation was a black-box technology. People knew what worked but not why. Enter Pasteur in the 1850s, who proved that yeast was a living organism, not a spontaneous chemical reaction. That revelation turned brewing into a microbiological practice and fermentation into an industry.

Today, fermentation is biotech on steroids. Breweries pitch genetically engineered yeast for higher alcohol tolerance or new flavor profiles. Distilleries use enzymes produced in industrial vats to maximize starch conversion. Cheese-makers inoculate milk with curated microbial cocktails to guarantee flavor.

Fermentation went from "leave it out and pray" to "design the genome, control the temp, log the data." Natural fermentation is messy, diverse, and unpredictable. Man-made fermentation is clean, consistent, and scalable. One gives you farmhouse funk; the other gives you Coors Light.

Why We Still Love the Wild Side

Here's the paradox: even with industrial precision, humans still chase wild fermentation. Craft beer enthusiasts value spontaneous fermentations, such as Belgian lambics, in which wild microbes enter through open windows. Sourdough enthusiasts boast about the "local flavor" of their starters, a microbial terroir you can't replicate. Kimchi lovers embrace the funk that shifts from batch to batch.

Why? Because chaos tastes good. Wild fermentation produces complex symphonies of compounds that man-made systems rarely replicate. Controlled fermentation is reliable. Wild fermentation is exciting. Both matter.

Fermentation and the Brain

Alcohol deserves its own sidebar. Ethanol is a small, polar molecule that easily crosses the blood-brain barrier. Once inside, it hijacks neurotransmitters: enhancing GABA (*relaxation*), dampening glutamate (*memory*), and releasing dopamine (*pleasure*). That's why alcohol feels warm and fuzzy and can make bad karaoke seem like a good idea.

Other fermentation products play tricks, too. Acids trigger salivation. Esters smell fruity. Amines and sulfur compounds produce funk. Fermentation isn't just preservation; it's sensory engineering, carried out by microbes for their own survival and hijacked by humans for flavor.

The Future of Fermentation

What's next? Precision fermentation, using engineered microbes to produce everything from vanilla to dairy proteins without cows. Yeast that secretes casein. Bacteria that brew omega-3 oils. Imagine cheese without cows, beer without barley, and coffee without beans. It's already happening.

At the same time, wild fermentation is resurging across culture and counterculture: natural wines, foraged yeasts, and home-fermented hot sauces. The tension between natural chaos and engineered control defines fermentation today, as it has for millennia.

Rot as Revelation

Fermentation is where nature and science shake hands. Wild microbes taught us that decay could be delicious. Human tinkering turned it into one of Earth's most powerful technologies. From beer halls to biotech labs, the principle remains the same: sugar in, transformation out.

The difference is scale and intention. Naturally occurring fermentation is a messy art. Man-made fermentation is a disciplined science. But both remind us that we're not the only cooks in the kitchen. Microbes have been running this show long before we arrived; we just learned to aim them.

So next time you sip wine, tear into sourdough, or fork into kimchi, remember: you're eating rot. Controlled, delicious, civilization-defining rot. It might be the most important experiment humanity has ever run.

Chapter Eight – Wheat and Corn: Twin Pillars of Civilization

Grains are more than food. They are cultural signatures, geographic markers, and economic engines. Few crops embody this truth more than wheat and corn. For millennia, they grew worlds apart, wheat across the Old World and corn across the New. Each fed empires, sustained rituals, and defined cuisines. Today, they grow side by side in fields across the American Midwest, as if history itself had been collapsed into a single agricultural panorama.

Their convergence was not inevitable. It was the product of trade, conquest, science, and globalization. Together, they show how deeply one kernel of starch can alter the fate of entire peoples.

The Staff of Life

Wheat's story begins in the Fertile Crescent around 10,000 years ago. The first domesticated varieties, einkorn and emmer, were unassuming grasses that carried a genetic gift: gluten.

Gluten, that maligned protein in modern diets, was wheat's superpower. When flour and water were combined and kneaded, the glutenin and gliadin proteins tangled into elastic networks, trapping air from wild yeast. What emerged wasn't porridge or flatbread; it was leavened bread. Light, airy, portable calories, perfectly matched to ovens and fire pits.

Bread gave civilizations a backbone. Wheat wasn't just sustenance; it was politics. Whoever controlled wheat controlled the city.

Wheat's chemistry underpinned its dominance. Its high protein content made it nutritionally robust. Its gluten networks made it versatile. And its grains, once milled, stored well, enabling granaries and, in turn, armies, cities, and bureaucracies. Wheat is not just a crop; it is the architecture of the state.

Corn: The American Giant

On the other side of the world, another grass was undergoing its own transformation. In the valleys of southern Mexico, about 9,000 years ago, indigenous farmers began selecting for larger kernels in a plant called teosinte. Over generations of careful cultivation, they coaxed it into a new species: maize.

Corn is biology, exaggerated. Each ear is a dense array of oversized seeds, each a starch-filled capsule wrapped in a tough pericarp. Unlike wheat, corn contains no gluten. It cannot rise into airy bread. Instead, it became tortillas, tamales, and porridges, foods designed for hand and hearth.

If wheat was the staff of life, corn was the communal table. It shaped Mesoamerican cosmology: the Popol Vuh, the Mayan creation story, declares that humans were formed from maize dough. Corn was not metaphorical; it was life itself.

Chemically, corn posed a problem. Its niacin (*vitamin B3*) was locked in forms the human body could not easily absorb. Cultures that relied solely on corn, without proper preparation, often developed pellagra, a deficiency disease. But Mesoamerican societies solved this with nixtamalization: cooking kernels with lime (*calcium hydroxide*). This simple, brilliant chemical trick released niacin, improved protein availability, and gave masa its distinctive flavor. It was applied chemistry masquerading as tradition.

Two Worlds Apart

For thousands of years, wheat and corn never met. Wheat shaped Old World economies, including Mediterranean empires, Middle Eastern dynasties, and European feudal systems. Corn, meanwhile, anchored New World civilizations, including the Maya, the Aztecs, and the Inca. Both grains scaled from subsistence to empire, yet each remained geographically isolated, separated by oceans and continents.

Then came 1492. The Columbian Exchange shattered agricultural boundaries. Wheat sailed west to the Americas, where Spanish colonists planted it in Mexico and the Andes. Corn sailed east, spreading across Europe, Africa, and Asia. Each grain colonized the other's hemisphere.

The results were transformative. Corn became a staple across Africa, fueling population booms. Wheat spread throughout the Americas, eventually dominating vast plains ideally suited to its cultivation. By the 19th century, the American Midwest had become the breadbasket of the world, though it was increasingly also the cornbasket.

Chemistry and Competition

Wheat and corn may share fields now, but they remain chemically distinct.

Wheat: high in gluten proteins, optimized for leavened bread, pasta, and baked goods. **Corn:** gluten-free, high in starch, suited for porridges, tortillas, and processed derivatives.

Corn, however, is more adaptable industrially. Grind it into meal, ferment it into ethanol, and refine it into high-fructose corn syrup. Its uses extend beyond food into fuel and sweeteners. Wheat, for all its cultural weight, is mostly food. Corn is both food and commodity feedstock, the raw material of modern industry.

Chemistry determines economics. Corn's starch granules break down easily into glucose, feeding microbial factories that produce syrups and alcohol. Wheat's gluten makes it indispensable for bread but less suitable for other industries. Together, they occupy distinct ecological niches in the global market.

Culture on the Plate

Even today, wheat and corn carry cultural identities. Wheat still evokes the Old World, with loaves of bread, pasta, and pastries. Corn evokes the New World, with tortillas, cornbread, and popcorn.

But their proximity has blurred traditions. In the United States, wheat bread and corn tortillas share supermarket shelves. Hybrid dishes, such as cornbread stuffing, wheat-flour tortillas, and sweet rolls glazed with corn syrup, show how globalization has fused two once-separate culinary universes.

Yet their symbolic roles linger. Bread retains its metaphorical weight: "daily bread," "breaking bread." Corn retains its earthy, communal symbolism, tied to land and indigenous heritage. They coexist without collapsing into sameness.

Global Stakes

Today, wheat and corn dominate global agriculture. Together, they cover hundreds of millions of hectares. Wheat remains a staple for over a third of the world's population. Corn surpasses wheat in sheer tonnage, with much of it diverted to animal feed, ethanol, and industrial uses.

Their coexistence is not without tension. Corn requires heavy fertilizer and water inputs; wheat is more tolerant of temperate conditions. Both rely on monoculture systems that invite pests and threaten biodiversity. Both shape geopolitics: Russian wheat exports ripple through global markets, and American corn subsidies reshape diets worldwide.

In the shadow of diverse climates, their stories converge. Breeding programs aim to develop drought-resistant wheat and pest-resistant corn. Genetic engineering introduces traits into both, blurring the line between traditional cultivation and biotech design. Once separated by oceans, they now share DNA in labs.

Two Grains, One World

Wheat and corn began as distant cousins, two grasses domesticated on opposite sides of the world. Each sustained empires and was encoded in myth and ritual. For millennia, they never touched. Now they grow side by side in Iowa fields, their kernels fueling everything from bread loaves to soda cans.

The irony is rich. Wheat, the Old World's gift of gluten and leavened bread, and corn, the New World's gift of masa and tortillas, are no longer cultural opposites. They are agricultural neighbors, with their destinies intertwined.

Together, they remind us that food is never just biology. It is culture, economy, chemistry, and history, all packed into kernels of starch. To eat wheat or corn is to consume not just calories but civilization itself.

Chapter Nine – The Politics and Economics

of Food Science

Here's the uncomfortable truth: food is never just food. It's policy, profit, and power disguised as calories. You think you're eating a bowl of cereal, but you're actually consuming the end result of centuries of subsidies, corporate lobbying, and scientific tinkering.

Food science doesn't live in a vacuum. It lives in boardrooms, legislatures, and the global marketplace. When chemistry meets economics, you get an experiment with billions of human subjects.

Subsidies: The Invisible Ingredient

Take corn. The U.S. grows about 15 billion bushels of it each year, blanketing the Midwest in endless seas of green. Why? Federal subsidies make it profitable even when the market doesn't. Those subsidies began during the Great Depression to stabilize farmers' incomes. They never really went away.

Enter food science. Cheap corn became cheap starch. Cheap starch became high-fructose corn syrup (*HFCS*). HFCS became the sweetener in everything from soda to ketchup. Congratulations: farm policy rewired the American diet.

The same story plays out with wheat, soybeans, and rice. Government programs don't just support farmers; they shape what scientists are paid to research, what corporations are paid to process, and ultimately what ends up on your plate. When the "free market" decides what you eat, it's usually a decision made with a heavy dose of government intervention.

Corporations as Chemists

First, corporations. A food scientist in a lab doesn't just dream up new products because it's fun. They're paid to solve corporate problems. How do we keep bread soft for 2 weeks on the shelf? How do we make low-fat ice cream taste like the full-fat version? How do we shave half a cent off each snack bar without losing customers?

Cue the army of additives. Emulsifiers keep oil and water from separating. Preservatives such as sorbates and benzoates help keep microbes at bay. Artificial flavors simulate strawberries without requiring a single strawberry. None of these molecules is "bad" in themselves; they're often clever chemistry. But their existence is tied to economics: scale, shelf life, and cost reduction.

And here's where politics enters: regulators must decide whether those additives are safe. Corporations lobby for approvals, fund studies, and sometimes bury inconvenient results. The "Generally Recognized as Safe" (*GRAS*) designation in the U.S. often relies on industry-funded research. Translation: the fox helps write the rules for the henhouse.

Nutrition Science: Conflicted by Design

Nutrition is supposed to be a science. In reality, it's often economics wearing a lab coat.

Case in point: the low-fat craze of the 1970s–1990s. U.S. dietary guidelines warned that fat was the cause of heart disease. The food industry pivoted to "low-fat" everything, including cookies, yogurts, and snack bars, all pumped full of sugar and starch to make them palatable. Rates of obesity and diabetes climbed.

Was fat the villain? Turns out, the story was more complicated. But decades of food science were bent to political consensus and industry convenience. The sugar industry, as we now know from internal documents, funded research that downplayed sugar's risks and emphasized fat instead. Scientists became pawns in a corporate endgame, and entire generations ate accordingly.

Food as Commodity, Not Meal

Here's a shift worth noting: food stopped being food and became a commodity. Wheat isn't bread. It's futures contracts on the Chicago Board of Trade. Soy isn't tofu. It's feed for livestock, oil for biofuels, and filler in processed foods. Corn is less about tortillas than about ethanol and sweeteners.

When food is a commodity, economics trumps taste or nutrition. A bumper crop of corn doesn't mean cheaper meals; it means lower futures prices that ripple through energy markets, animal feed, and international trade. A drought in Ukraine doesn't just hurt local farmers; it spikes global wheat prices, destabilizing governments in North Africa. Bread riots are often more about commodity futures than about bakers.

Food science in this context isn't just about making things edible; it's also about understanding the science behind food. It's about making commodities flexible. Soy proteins are used to create fake meat. Corn starch is broken down into glucose, which is then fermented into HFCS or ethanol. Wheat is fractionated into gluten for baking, starch for industry, and bran for supplements. Every kernel is dissected into chemical profit centers.

Global Trade: Calories as Diplomacy

Food is a form of political leverage. The U.S. exports corn and soybeans to China as feed for China's pigs. Brazil clears rainforests for soy plantations to meet global demand. Russia and Ukraine's wheat exports are bargaining chips in geopolitical crises.

Food science becomes diplomacy here. Fortified grains, enriched with iron, folic acid, and vitamin D, are tools of international aid. Instant noodles, engineered for shelf stability, feed refugees in camps. Golden Rice, genetically engineered to produce beta-carotene, was designed to combat vitamin A deficiency in developing nations.

But each of these is wrapped in politics. Genetically modified organisms (*GMOs*) spark battles over labeling and regulation. Food aid can flood local markets, undercutting local farmers. Trade wars often begin with tariffs on agricultural products, such as soybeans and corn. Calories become currency.

The Double-Edged Sword of Technology

Food science promises solutions, including fortified cereals to fight malnutrition, lab-grown meat to reduce carbon emissions, and precision fermentation to replace dairy cows. Each breakthrough comes with glowing headlines about saving the planet.

And sometimes, those promises deliver. Iodized salt virtually eradicated goiter in many countries. Pasteurization slashed deaths from milk-borne diseases. Refrigeration revolutionized preservation.

But sometimes, the fixes create new problems. Processed convenience foods solved postwar hunger and time scarcity, then fueled epidemics of obesity and diabetes. High-yield monocultures boosted production, then drained soils and poisoned rivers through fertilizer runoff. Lab-grown meat may cut methane emissions, but it raises questions about energy use, patents, and ownership of food.

Food science is never neutral. Every innovation is a trade-off among health, environment, profit, and power.

The Consumer Paradox

Here's the paradox: we, the eaters, think we're in control. We select items from shelves, scroll through menus, and choose brands. Yet our choices are shaped upstream by policies, subsidies, corporate strategies, and scientific decisions we often never see.

We "like" sugar because our biology does. We also like it because it's cheap, abundant, and added to almost every processed food. We buy meat because it's affordable, but only because feed crops are subsidized, antibiotics keep animals alive in crowded barns, and scientists have designed additives to extend shelf life.

Consumers are the test subjects in a vast experiment conducted jointly by labs, boardrooms, and legislatures.

Who Owns Dinner?

So what's next? Precision fermentation may produce milk proteins without cows. CRISPR may be used to edit crops for drought resistance. Vertical farms may churn out greens in skyscrapers. Lab-grown meat may redefine "livestock."

But the real question isn't whether we can. It's who profits? If patents lock up engineered seeds, if biotech firms own the microbes that brew milk proteins, and if carbon credits turn farms into financial instruments, then food science becomes less about feeding people and more about feeding markets.

We're at a crossroads. Do we harness food science to build resilience and equity? Or double down on commodification and control? The chemistry is neutral. Politics and economics are anything but.

The Hidden Recipe

Every bite you eat contains more than calories and flavor. It reflects policy decisions, corporate strategies, and scientific trade-offs. It is subsidized, engineered, and marketed long before it reaches your plate.

Food science is extraordinary. It made bread softer, milk safer, beer cleaner, and vegetables last longer. But it also made soda cheaper than water, candy bars ubiquitous, and monocultures dangerous. It solved hunger in one place while creating chronic disease in another.

The hidden recipe for modern food is one part chemistry, one part economics, and one part politics. Ignore any of those ingredients, and you miss the real flavor of dinner.

So the next time you tear open a bag of chips or sip a soda, remember: you're not just eating food. You're eating the outcome of a global experiment in policy, profit, and science. And you didn't sign the consent form.

Chapter Ten – The Secret Science of Cooking

They Don't Want You to Know

Okay, I'll admit: "secret science" sounds like clickbait. Like I'm about to sell you a PDF titled "10 Kitchen Hacks Big Sugar Hates!" But here's the thing: there are tweaks that professional kitchens, industrial food scientists, and Michelin chefs use every day that most home cooks don't know about. They're not mystical. They're chemistry, physics, and a little economics, hidden in plain sight.

Let's crack them open.

***

Artifice One: Salt Is a Time Machine

Everyone knows salt makes food taste better. But here's the secret: salt isn't just a seasoning. It's a molecular time-traveler.

When you salt meat hours before cooking, sodium and chloride ions seep into muscle fibers. They loosen protein structures, helping water stay trapped during cooking. The result? Juicier steak, tastier chicken, more forgiving pork chops.

Chefs call this "dry brining." Scientists call it osmotic equilibrium and protein denaturation. Either way, it delivers the same effect as expensive marinades, without the mess.

But here's the kicker: salt also enhances flavor perception. It suppresses bitterness and amplifies sweetness and umami. That's why chocolate chip cookies taste better with a sprinkle of flaky salt. It's not fashion; it's sensory biology. The pros know this. Now you do, too.

***

Artifice Two: Rest Is Cooking

Professional chefs know this, but most home cooks don't: pulling meat off the heat isn't the end of cooking. It's the middle.

Here's the physics: heat moves from the outer layers of meat toward the center. When you pull a steak at an internal temperature of 130°F, the outer crust is hotter, closer to 400°F. Take it off the grill, and the heat continues migrating inward. That "carryover cooking" can raise the internal temperature by 5–10°F.

That's why chefs always "rest" meat before slicing. It's not superstition. It's thermal redistribution. Juices redistribute, fibers relax, and temperature stabilizes. Slice too early, and you lose flavor to the cutting board. Wait a few minutes, and you get steak engineered for tenderness.

The secret? Cooking doesn't stop when you stop applying heat. It stops when equilibrium dictates.

***

Artifice Three: Acids Are Magic Wands

Ever wonder why vinaigrette brightens a salad or why a squeeze of lemon transforms grilled fish? Acids are flavor multipliers.

Chemically, acids donate protons (H^+ *ions*). On the tongue, that proton surge excites taste receptors, amplifying the perception of other flavors. That's why ceviche works: lime juice denatures fish proteins, changing texture without heat. Same reaction as cooking, different input.

Chefs use this constantly. Tomato paste adds acidity to cut through fat. Vinegar cuts through heaviness. Even desserts benefit; fruit pies sing when a little lemon juice tweaks the sugar-fat balance.

The secret pros know: if your dish tastes "flat," it doesn't need more salt. It needs acid.

***

Artifice Four: Brown Is Flavor

This one's almost criminal in its secrecy. The brown crust on steak, toast, or roasted veggies isn't just "cooked." It's the Maillard reaction: a complex network of reactions between amino acids and sugars that begins around 300°F.

The result? Hundreds of volatile compounds that smell like nuts, caramel, toast, or umami.

Professional kitchens manipulate the Maillard reaction with precision. They pat steaks dry to remove water (*which caps the surface temperature at 212°F*). They preheat pans until they're practically glowing. They know: no brown, no flavor.

Your toaster? Same principle. Bread, heat, and amino acids equal breakfast perfume. The secret isn't really a secret; it's chemistry disguised as comfort.

***

Artifice Five: Emulsions Are Controlled Chaos

Mayonnaise looks boring, but it's scientific sorcery: oil and water, sworn enemies, suspended together in harmony.

How? Emulsifiers. In mayo, lecithin molecules from egg yolks act as molecular diplomats, with one end bonding to oil and the other to water. Whisk hard enough, and billions of droplets suspend in creamy stability.

Professional chefs apply this trick to sauces, dressings, and even cocktails. Hollandaise? An emulsion. Chocolate ganache? Same. Ice cream? Fat globules suspended in frozen sugar water.

The secret isn't "technique." It's emulsifiers and physics. Get the ratio right, and you can suspend almost anything in any other substance.

***

Artifice Six: Cold Is a Flavor Weapon

Freezing doesn't just preserve food; it changes its chemistry.

Take ice cream. Sugar lowers the freezing point of water, preventing the formation of large crystals. Churn while freezing to incorporate air (*overrun*), giving a creamy texture. Industrial freezers can drop temperatures so quickly that they generate micro-crystals, producing smoother ice cream than most home machines can achieve.

Even water benefits from the cold trick. Ever wonder why restaurant ice cubes are clear? They're frozen slowly from one direction, which forces air bubbles out. Clear ice doesn't just look fancy; it melts more slowly, diluting your cocktail less.

The secret: freezing is as much engineering as preservation. Temperature curves matter as much as recipes do.

***

Artifice Seven: Time + Microbes = Flavor Bombs

We covered fermentation earlier, but here's the "secret" angle: professional kitchens use microbes like sculptors use chisels.

Sourdough? It's wild yeast and lactic acid bacteria. Cheese? Mold and enzymes. Soy sauce? Fungi, salt, and months of patience. These aren't accidents. They're microbial ecosystems, engineered by tradition and refined by science.

The hidden trick: you're not always cooking with fire. Sometimes, you're cooking with time. Letting microbes do the work slowly and invisibly, creating flavors no stove could replicate.

Experts know when to step back and let nature take the load.

***

Artifice Eight: MSG and Umami

Ah, yes, the villainized hero. Monosodium glutamate.

MSG is the purest form of umami, the savory taste of glutamate ions. Tomatoes, Parmesan, and mushrooms all naturally contain it. MSG is just a concentrated sprinkle. It binds to receptors on the tongue (*the umami-specific T1R1/T1R3 complex*), enhancing depth and making everything taste meatier.

Why the controversy? Racism and bad science. A 1969 letter to a medical journal coined the term "Chinese Restaurant Syndrome," sparking decades of suspicion. Modern studies show that MSG is safe. In fact, many processed foods use it quietly because once you realize umami is addictive, why wouldn't you?

Edwards   Controlled Burn   110

The "secret" pros know: MSG isn't cheating. It's seasoning on a molecular level.

***

Artifice Nine: Texture Is Half the Flavor

Taste is only half the story. Texture is the other half, and chefs manipulate it as engineers do.

Want crunch? Fry food to dehydrate the surface into a brittle matrix. Want tenderness? Cook low and slow to convert collagen into gelatin. Want creaminess? Emulsify fats into microdroplets that coat the tongue.

The secret: flavor perception is multisensory. The crackle of a chip isn't just sound; it's part of taste. The jiggle of custard, the chew of al dente pasta, and the fizz of carbonation are all sensory hacks rooted in physics.

Chefs don't just cook for the tongue. They cook for the ears, eyes, and even memory.

***

Artifice Ten: Heat Isn't Just Temperature

Professional cooks don't just say "cook at 400°F." They care about *heat transfer.*

Conduction: Pan to steak. Direct, fast, intense. **Convection:** Hot air circulating in an oven. Slower, even. **Radiation:** Infrared from a broiler or the sun. Surface-level blasting.

Understanding the difference explains why a pizza stone produces a better crust (*thermal mass plus conduction*) or why convection ovens bake faster (*hot air transfers heat more efficiently than still air*).

The secret is not just "how hot," but "how heat moves."

***

Artifice Eleven: Flavor Pairing Algorithms

Here's one from the cutting edge. Food scientists now use chemical analysis to map volatile compounds in ingredients. Garlic, coffee, and chocolate, for instance, share specific sulfur compounds. That's why mole sauce works.

Chefs like Heston Blumenthal use this science to invent wild pairings, such as white chocolate with caviar and strawberries with balsamic. On paper, it looks insane. On a gas chromatograph, it's destiny.

The secret pros know: flavor isn't guesswork. It's data-driven matchmaking.

Why Don't We Hear About This?

It's not that experts are hiding things. It's that most of us don't ask, and most kitchens don't teach diners chemistry. Salt is "seasoning," not ion diffusion. Browning is "flavor," not Maillard. Freezing is "cold," not cryogenic crystallography.

But once you know the science, you see the strings behind the puppet show. The magic isn't gone... It's richer.

***

The Curtain Pulled Back

The secret science of cooking is this: nothing is magic. Every trick is atoms moving, bonds breaking, and molecules rearranging. Professionals know it, use it, and sometimes guard it with mystique. But you don't need to be a Michelin-starred chef or a food scientist to understand it.

Salt is chemistry. Browning is chemistry: emulsions, fermentation, freezing, all of which are chemistry. The "secrets" aren't really secrets at all. They're tools. Once you see them, you can use them, too.

Cooking is an applied science that tastes good. The experts may not say that outright, but now you know: the kitchen is a lab, and dinner is the experiment.

Chapter Eleven – Kitchen Myths: Science vs. Folklore

The kitchen is full of myths, little bits of folklore passed down like family heirlooms. Some are harmless, like "grandma's secret ingredient is love." (*It isn't; it's usually butter.*) Others waste your time, rob your food of flavor, or actively make your cooking worse.

It's time to put these myths under the microscope. Literally. Once you look at what's actually happening at the molecular level, many "rules" fall apart like overcooked stew meat.

***

Myth One: "Searing Meat Seals in the Juices"

Nope. Not even close.

Here's the claim: blasting a steak at high heat "seals" the outside, creating a barrier that locks in juices. Sounds plausible. Looks dramatic. Smells amazing. Totally false.

What searing actually does is kick off the Maillard reaction, a cascade in which amino acids and sugars break down into hundreds of new flavor molecules at around 300°F. You're creating flavor, not waterproofing. In fact, moisture loss is about the same whether you sear first, last, or not at all.

The science: meat is about 75% water. That water is inside cells. Heat ruptures those cells, and the water escapes. No amount of crust is going to stop physics.

The truth: searing is still great because browned meat tastes like heaven. But the "juices sealed in" claim? Pure marketing from the 19th century, when German chemist Justus von Liebig misinterpreted his own experiments. The myth spread, stuck, and sizzled into eternity.

***

Myth Two: "Alcohol Cooks Off Completely"

Sorry to break it to you, but that wine in your pan sauce isn't just a flavor memory. Some of it is still there.

Yes, alcohol boils at about 173°F, lower than water's 212°F. But in a mixture, evaporation is slower and less complete. USDA studies show that even after simmering for two hours, some alcohol remains. After flambéing (*setting it on fire*), you might still have 75% left.

So if you cook with booze, you're still serving booze. Not a ton, but enough that "don't worry, it all burns off" is scientifically inaccurate.

The truth: alcohol contributes more than a buzz. It dissolves flavor compounds that water and fat can't, unlocking aromas in herbs, spices, and tomatoes. That's why wine sauces have a complex flavor. Don't use cooking wine as an excuse to drink straight from the pan, but don't assume it vanishes, either.

***

Myth Three: "Microwaves Destroy Nutrients"

Ah, the microwave, our most maligned appliance.

The myth says that nuking food "kills nutrients." The reality? Microwaving often preserves them better. Here's why: nutrient loss usually results from prolonged heat exposure or leaching into water. Boil broccoli for 15 minutes, and you'll lose half the vitamin C to the cooking water. Microwave it with a splash of water for two minutes, and you'll keep most of the good stuff.

The science: microwaves excite water molecules, heating food from the inside out. The energy isn't ionizing (*no DNA damage, no mutant carrots*). It's just fast. Less time and less water mean more nutrients.

So no, microwaves don't "destroy food." They're misunderstood physics in a beige box.

***

Myth Four: "You Have to Rinse Pasta"

Only if you hate yourself.

The myth: Rinse pasta after boiling to "stop cooking" and prevent clumping. The problem? Rinsing washes away the surface starch, the glue that helps the sauce adhere. You end up with slippery noodles that repel tomato sauce like Teflon.

Professional kitchens don't rinse (*unless it's for cold pasta salad*). They save some of that starchy pasta water to emulsify sauces, creating a glossy, clingy texture.

So, unless your goal is sadness in a bowl, skip the rinse. Sauce loves starch.

***

Myth Five: "Cooking Removes All Bacteria"

If only.

Cooking kills most bacteria when temperatures reach high enough levels, but not all bacterial toxins. Some, like those from Staphylococcus aureus, can survive heat. Others, like spores from Clostridium botulinum, laugh at boiling water. That's why pressure canning exists. You need temps above 240°F to neutralize them.

The takeaway: cooking is a safety tool, not a magic bullet. Proper storage, handling, and reheating matter just as much. The myth persists because "kill it with heat" sounds reassuring. The truth is messier and more scientific.

***

Myth Six: "Cast Iron Is Fragile"

You've heard it: "Never use soap on cast iron! Never cook tomatoes in it! Don't let water touch it!"

Nonsense. Cast iron is tough. It's iron. You can drop it, scrub it, or even cook acidic sauces in it. Worst case, you strip some seasoning, a thin layer of polymerized oil. Guess what? You can re-season it by heating more oil until it polymerizes again. It's self-healing cookware.

The myth survives because cast iron requires care (*avoid leaving it wet*). But fragile? Please. Cast-iron pans outlive owners.

***

Myth Seven: "You Need Fancy Knives"

This one's a favorite among knife companies. Yes, a $300 Damascus steel blade is gorgeous, but sharpness depends on angle and maintenance, not price.

A $30 chef's knife, properly sharpened, outperforms a dull luxury knife every time. The science is simple: sharp edges reduce the pressure required to sever fibers, preserving texture and flavor. Dull edges crush cells, releasing enzymes and juices.

What experts know (*and marketers hide*): honing and sharpening matter more than metallurgy. The fancy knives are fun, but science says function is about geometry and maintenance.

***

Myth Eight: "Pink Pork Means Undercooked Pork"

Not anymore.

For decades, pork was overcooked into gray leather because of fears of trichinosis. But trichinella parasites have been virtually eliminated in commercial pork in many countries. In 2011, the USDA lowered the safe internal temperature from 160°F to 145°F. That means a blush of pink is fine... as long as it reaches 145°F and rests for three minutes.

Science: it's about internal temperature, not color. Meat color depends on myoglobin, which varies by animal and muscle. Pink pork is safe pork if the thermometer agrees.

***

Myth Nine: "Marinating Meat Tenderizes It Deeply"

Marinades are flavor on the surface, not magic potions.

Acidic marinades (*such as vinegar, citrus, or yogurt*) denature proteins only in the outer few millimeters. They don't penetrate deeply. Oil-based marinades mostly add surface flavor. Enzyme-based marinades (*such as papaya and pineapple*) break down proteins more aggressively, but mainly at the surface.

If you want deep tenderness, you need time and low heat (*braising*) or mechanical disruption (*such as pounding*). Marinades equal flavor paint. Slow cooking equals structural engineering.

***

Myth Ten: "More Heat = Faster Cooking."

Yes, water boils faster on high heat. No, that doesn't mean your food cooks faster once the water is boiling. Pasta cooks at 212°F, whether the flame is medium or full blast. Stew simmers the same way. Cranking the burner wastes energy and risks boil-overs.

The science: once water boils, it can't get hotter without pressure. More heat just makes bubbles more violent, not the temperature higher. The myth persists because aggressive bubbles look like progress.

***

Myth Eleven: "Spicy Food Burns Your Tongue"

Not technically. Capsaicin in chili peppers binds to TRPV1 receptors, the same receptors that detect heat. Your brain interprets that binding as "burning," even though no actual heat damage occurs (*unless you rub your eyes*).

This molecular mimicry trick is why spicy food feels hot without scalding tissue. Your tongue is basically running bad software. The pros know this and use capsaicin levels like volume knobs for cuisine.

***

Myth Twelve: "Bread Goes Stale from Drying Out"

Counterintuitive but true: staling isn't about water leaving bread. It's about starch retrogradation, in which amylopectin molecules crystallize and squeeze water into less available forms. Refrigeration accelerates this process, which is why bread goes stale faster in the fridge than on the counter.

The fix? Reheating breaks up the crystals, making bread soft again (*temporarily*). The toaster isn't just warming bread; it's reversing the chemistry of starch.

***

Conclusion: Myths as Comfort Food

Why do these myths persist? They make cooking feel simpler, safer, and more predictable. "Searing seals juices" sounds reassuring. "Alcohol cooks off," comforts parents. "Cast iron is fragile" makes ownership feel mystical.

But science doesn't care about comfort. It cares about molecules, reactions, and physics. The good news? When you understand the truth, you don't lose the magic; you gain control.

Cooking myths are like lab superstitions. Bust them, and the kitchen becomes clearer, smarter, and frankly more fun.

So go forth with thermometers, acids, emulsifiers, and microwaves. Cook with science, not folklore. And if Grandma insists her way is better, just smile, nod, and quietly know you're running a tighter experiment.

Chapter Twelve – Preserving Food:

Holding Back Time

If cooking is the science of transformation, food preservation is the science of defiance. To preserve is to push back against entropy, cheat decay, and wrestle with time itself. Every loaf of bread, every cut of meat, every jar of jam shares the same fate if left alone: microbes move in, enzymes unravel molecules, and nature insists that what was once alive must return to the soil.

Preservation is humanity's way of saying, "not yet."

Throughout history, every culture developed its own preservation methods, including salting, smoking, fermenting, freezing, and canning. Each method reflects both science and geography: how people harnessed chemistry to keep food edible. Today, in supermarkets where strawberries sit plump in January, and fish is shipped across continents, preservation has become invisible. But behind the curtain are the same tricks, refined with modern tools.

Let's unpack the do's and don'ts, the tools and pitfalls, of keeping food at bay.

***

The Science of Decay

To preserve food, you first need to understand what you're fighting.

Two enemies dominate the battlefield: **microbes** and **enzymes**.

Microbes, including bacteria, yeasts, and molds, see food as fuel. They consume sugars, proteins, and fats, leaving behind acids, gases, and toxins. Sometimes we harness them (*fermentation*), and sometimes we fight them.

Enzymes are the food's own proteins, tiny chemical scissors still at work after harvest or slaughter. They brown apples, soften vegetables, and break down tissues.

Preservation works by slowing or stopping these forces. Remove water, alter pH, raise or lower temperature, or starve microbes of oxygen. Every method is an intervention against entropy.

***

Drying: The First Technology

Drying is the oldest trick in the book. Ancient peoples laid fish on rocks, strung herbs in the sun, and spread grain on mats. By reducing water content, they created inhospitable conditions for bacteria. Water is life; take it away, and microbes can't thrive.

Today, dehydration is used to make jerky, dried fruit, and powdered milk. Freeze-drying goes further: it freezes food, then sublimates the ice directly into vapor under vacuum. Astronauts eat crunchy freeze-dried strawberries because NASA perfected what Neolithic herders discovered by accident: less water, more time.

Do: keep air moving, use low, steady heat, and store in airtight containers. **Don't:** assume dried food is immortal. Humidity is its Achilles' heel.

***

Salting and Sugaring: Osmosis as a Weapon

Salt and sugar are preserved by osmosis, drawing water from microbial cells until they shrivel. Salted fish fueled sailors on transoceanic voyages. Sugared fruits became jams, jewels of sweetness sealed in jars.

The chemistry is elegant: solutes lower water activity (a_w). Bacteria require a minimum a_w to grow; drop below that, and their metabolism grinds to a halt. That's why honey, nearly pure sugar, never spoils. Archaeologists have found honey in Egyptian tombs that is still edible after 3,000 years.

Do: measure carefully. Ratios matter; half-salted meat is an invitation to botulism. **Don't:** trust appearances. Salted food may look fine, but it can harbor dangerous microbes if stored improperly.

***

Pickling and Acidification

Pickling is a preservation method that relies on acidity. Vinegar (*acetic acid*) lowers the pH below 4.6, the threshold at which Clostridium botulinum, the bacterium responsible for deadly botulism, can't produce toxins. Brining vegetables in vinegar creates inhospitable conditions for most microbes.

Fermented pickles, sauerkraut, and kimchi develop acidity through fermentation. Lactic acid bacteria convert sugars into acids, souring the environment and adding layers of flavor.

Do: keep vegetables submerged. Oxygen exposure invites mold. **Don't:** rely on guesswork. Botulism is rare but unforgiving.

***

Cold: From Icehouses to Freezers

Cold slows chemical reactions. For centuries, icehouses stored blocks of winter ice under sawdust, extending their life into summer. Refrigeration democratized cold, first with iceboxes and then with mechanical refrigerators. Suddenly, milk didn't sour overnight. Meat lasted a week instead of a day.

Freezing goes further: at 0°F, microbial activity virtually stops. Enzymes slow dramatically. Fish caught in Alaska can arrive in New York tasting fresh. But freezing doesn't halt decay forever; it only delays it. Ice crystals rupture cell walls, and thawing causes leaks that degrade the texture.

Do: freeze quickly to minimize ice crystal size (*flash freezing is best*). **Don't:** refreeze repeatedly. Each thaw-refreeze cycle damages texture and increases the risk of microbial contamination.

***

Heat: Canning and Sterilization

The 19th century brought us one of preservation's great revolutions: canning. Nicolas Appert, a French confectioner, discovered that sealing food in jars and heating them prevented spoilage. He didn't know why (*Pasteur hadn't yet linked microbes to decay*), but it worked.

Modern canning uses pressure and heat to sterilize contents. Low-acid foods (*meat, beans*) require pressure canners to reach

240°F, which kills botulism spores. High-acid foods (*such as tomatoes and fruits*) can be safely water-bath canned.

Do: follow tested recipes. Home canning is chemistry with stakes. **Don't:** improvise under time pressure. "Rustic charm" isn't worth botulism.

***

Smoke: Flavor Meets Preservation

Smoking was originally about survival, not barbecue competitions. Wood smoke contains antimicrobial phenols and aldehydes. Together with drying and salting, it extended the shelf life of fish and meat.

Today, we mostly smoke for flavor. Cold smoking (*below 90°F*) preserves; hot smoking (*above 200°F*) cooks and flavors at the same time. Both rely on the chemistry of wood combustion, in which lignin breaks down into guaiacol and syringol, those smoky aromatics we humans find irresistible.

Do: combine smoke with salt or refrigeration for safety.

Don't: assume smoke alone preserves indefinitely.

***

The Pitfalls of Preservation

Here's where things go wrong:

The "smell test." Many pathogens don't announce themselves with odor. Botulinum toxin is odorless. Salmonella doesn't always stink. Trust science, not your nose.

Improvising ratios. Preservation depends on precise thresholds of salt, sugar, acid, or temperature. A "pinch" too little can create a microbial paradise.

Ignoring storage. A sealed jar left in sunlight will degrade. Plastic bags allow moisture in. Preservation is only as strong as storage.

Preservation isn't superstition. It's applied chemistry, and chemistry has rules.

***

Modern Tools

Preservation today is both humble and high-tech.

Vacuum sealers remove oxygen, slowing oxidation and freezer burn.

Desiccant packets fight humidity in dried foods.

Refrigerated transport turns global trade into an everyday miracle, with Argentine beef, Norwegian salmon, and California strawberries appearing thousands of miles from their origins.

Food additives, such as ascorbic acid, nitrates, and BHT, extend shelf life further, sometimes controversially.

The boundary between traditional and modern is blurrier than it seems. Vacuum sealing is just fancy drying. Ascorbic acid is just vitamin C in a lab coat. Preservation has always been about pushing back decay with whatever tools we can wield.

***

Preservation and Culture

Every preservation method carries culture with it. Smoked hams in Europe, kimchi crocks in Korea, salted cod in the Caribbean, and pickled cucumbers in Eastern Europe were all survival strategies that became hallmarks of identity.

Preservation shapes cuisines not only by what lasts but also by how it transforms. Salted cod tastes nothing like fresh fish. Sauerkraut isn't raw cabbage stretched out; it's a new food, born of time and microbes. In this way, preservation is not only practical but also creative, turning scarcity into flavor.

***

The Hidden Costs

Preservation has downsides.

Industrial preservation, canning, refrigeration, and additives provided abundance but disconnected us from seasonality. Strawberries in January are possible, but they taste like sugar water compared to June's field strawberries.

Preservatives extend shelf life but can raise new health questions. Nitrates protect against botulism in cured meats but may be linked to cancer risk.

Refrigeration reshaped trade patterns, favoring large-scale global systems over local food economies.

Every preservation method trades flavor, texture, or ecology for time.

***

Preservation as Defiance

Yet preservation is also resilience. A jar of jam on the shelf, a freezer full of meat, a sack of dried beans... these are shields against uncertainty. They represent foresight, planning, and sometimes even survival itself. In wars, famines, and pandemics, preservation is the difference between hunger and sustenance.

At its best, it's joy. Few pleasures match cracking open a jar of summer peaches in midwinter or slicing into smoked cheese aged for months. Preservation lets us eat not just calories but memories.

***

Conclusion: Holding Back Time

Preservation is not about perfection. Nothing halts decay forever. It's about buying time, stretching abundance through lean months, stretching geography across oceans, and stretching seasons into memories.

The do's and don'ts, the tools and pitfalls, are science's way of codifying what generations already knew: preserving food is preserving ourselves.

Whether you salt a ham, freeze a steak, can tomatoes, or vacuum-seal leftovers, remember: you're not just storing calories. You're bending time, shaping culture, and keeping chaos at bay, one jar, one freezer, one smoky log at a time.

Chapter Thirteen – Kitchen Experiments:

Food Science You Can Play With

Here's the truth: cooking is already a series of experiments. You combine chemicals (*ingredients*), apply heat or pressure (*energy*), and wait for molecules to rearrange. The difference between "dinner" and "science experiment" is usually whether you eat the results.

But some experiments are too fun not to treat like science fair projects. The best part? You don't need a lab coat or fancy gear: just your kitchen, a few everyday ingredients, and the willingness to get messy. However, if you have lab coats available, I highly encourage you to wear them in the kitchen for these experiments or even during your daily kitchen routines.

***

Experiment One: Rock Candy – Crystal Chemistry

You Can Eat

What you need: sugar, water, a saucepan, a glass, a string, or a popsicle stick.

Boil sugar to make a supersaturated solution (*basically, dissolve more sugar than the water can normally hold*). Suspend a string in it and wait. Over a few days, sugar molecules lock together into crystals big enough to sparkle.

The science: crystallization is about molecules finding order. Each sugar crystal is like a Lego tower, with glucose and fructose molecules stacking in neat, repeating patterns. Once it starts, it snowballs. New molecules stick to existing ones, and the crystal grows.

The fun: you've made candy by literally watching chemistry build itself. Bonus points for food coloring, because nothing says science like neon sugar.

***

Experiment Two: Bag Ice Cream – Freezing Point Depression in Action

What you need: milk, sugar, vanilla, a small zip bag, a big zip bag, ice, and salt.

Mix the milk, sugar, and vanilla in the small bag. Place it inside the big bag filled with ice and a hefty dose of salt. Shake as if your life depends on it. Ten minutes later: ice cream.

The science: salt lowers the freezing point of water. Instead of freezing at 32°F, salty ice water can drop to 20°F. That colder brine pulls heat from your milk mixture faster, freezing it into ice cream. It's thermodynamics in disguise as dessert.

The fun: it's basically child labor with delicious payoffs. You'll feel like a wizard conjuring ice cream from nowhere.

***

Experiment Three: Homemade Soda – Yeast as a Carbonation Machine

What you need: juice or sweetened tea, yeast, and a clean plastic bottle.

Pour the sweet liquid (*juice works well here*) into the bottle, sprinkle in a pinch of yeast, seal it, and let it sit at room temperature for 12–24 hours. Open carefully. Fizzy soda.

The science: yeast eats sugar, burps CO_2, and produces a tiny amount of alcohol. In a sealed bottle, the gas dissolves into the liquid. Open it, and Henry's Law takes over. The dissolved gas escapes, forming bubbles.

The fun: you've just built a soda factory powered by microbes. Just don't let it go on too long, or you'll accidentally invent questionable wine.

***

Experiment Four: The Egg in a Bottle – Pressure at Work

What you need: a hard-boiled egg, a glass bottle with a narrow neck, paper, and matches (*or a lighter*).

Peel the egg, light a piece of paper, drop it into the bottle, and quickly place the egg on top. As the flame goes out, the egg is sucked into the bottle like magic.

The science: burning paper heats the air, expanding it. The flame goes out, the air cools, and the pressure inside drops. Higher outside pressure pushes the egg in. It's gas laws in action.

The fun: it looks like witchcraft. Bonus points if you pretend you're summoning protein demons.

***

Experiment Five: Popcorn – Steam Power in Every Kernel

What you need: popcorn kernels, a pan, and a lid.

Heat kernels until, pop! Pop! Pop!

The science: each kernel has a tiny water droplet trapped inside a starchy shell. Heat turns the droplet into steam, and pressure builds until the shell bursts. The starch gelatinizes, expands, and cools into a crunchy foam.

The fun: you're making edible explosions. Plus, you can explain to your kids that each piece of popcorn is a miniature steam engine.

***

Experiment Six: Invisible Ink Lemon Juice – Acids and Oxidation

What you need: lemon juice, paper, a lamp, or an iron.

Write a message with lemon juice, then let it dry. Heat the paper, and the message turns brown.

The science: lemon juice weakens cellulose fibers in paper. Heat accelerates oxidation, browning the weakened spots first.

The fun: secret spy notes powered by chemistry. Bonus points if you whisper "For science!" while revealing your message.

***

Experiment Seven: The Floating Egg – Density Games

What you need: eggs, water, and salt.

Place an egg in plain water. It sinks. Stir salt into the water until it floats.

The science: saltwater is denser than freshwater. When the solution's density exceeds the egg's, buoyancy lifts the egg.

The fun: it's Archimedes' principle with breakfast food.

***

Experiment Eight: Whipped Cream – Trapping Air in Fat

What you need: heavy cream, sugar, and a whisk.

Whip until soft peaks form.

The science: fat molecules stabilize air bubbles. Whisking unfolds proteins, trapping bubbles in a network. You've created a foam, a gas dispersed in liquid fat.

The fun: real-time transformation. Liquid becomes fluff. Bonus: the payoff is strawberries and whipped cream, not just data.

***

Experiment Nine: Caramel – Controlled Pyrolysis

What you need: sugar, a saucepan, and a bit of patience.

Heat the sugar until it liquefies and turns golden brown. Don't walk away.

The science: caramelization is sugar molecules breaking apart and recombining into hundreds of new compounds. Sweetness turns bitter, floral, nutty, and smoky. Temperature control is everything. Go too far, and you've got blackened glass shards.

The fun: it feels dangerous and alchemical. Plus, you can pour it into shapes, letting science double as edible sculpture.

***

Do's and Don'ts for Kitchen Science

Do embrace the mess. Science is messy; kitchens were built to be cleaned.

Do talk through the "why" with kids and friends. The fun is in discovery, not just in the snack.

Don't confuse food science with safety shortcuts. Fermentation is great; botulism is not. Stick to safe experiments only.

Don't expect perfection. Science thrives on mistakes, and so do pancakes.

***

Why These Experiments Matter

They're fun, yes. But they also reframe the kitchen. Instead of a place for rote recipes, it becomes a laboratory of curiosity. Kids (*and adults*) see that salt isn't just seasoning; it's osmosis. Popcorn isn't just a snack; it's a pressure vessel. Whipped cream isn't just a dessert; it's a foam engineered by fat and protein.

The best way to understand food science isn't to read about it. It's to eat it.

***

Conclusion: Science at the Table

Food experiments don't just teach chemistry; they teach humility. You discover how much of cooking is controlled chaos and how many recipes are experiments refined over generations.

And you also discover joy: the squeal of soda fizzing, the grin as ice cream forms, the awe as an egg gets sucked into a bottle. These aren't just demonstrations; they're reminders that science is playful, edible, and all around us.

So go ahead: spill some sugar, burn some caramel, and fizz up some juice. Run the experiment, taste the results, and laugh at the failures. That's cooking, that's science, that's life.

Chapter Fourteen – The Kitchen Manifesto:

Food *Is* Science

Let's stop pretending. Cooking isn't an art with "a dash of science." It's a science with a dash of art. Every recipe is an experiment. Every meal is data. Every cook is a scientist, whether they admit it or not.

That bread you pull from the oven? Controlled fermentation and thermal polymerization. That pasta sauce? Acid-base balance with emulsions. That steak? Heat transfer and protein denaturation, accompanied by Maillard chemistry.

It's not magic. It's not instinct. It's molecules moving, colliding, and rearranging. The sooner we admit that, the sooner we can stop blindly following recipes and start deliberately running experiments.

***

Experiment, Eat, Repeat

Science is a discipline of curiosity, methodically pursued. Cooking is hunger, disciplined by fire. Combine them, and you get a kitchen lab that doubles as a dining room.

Meal planning? That's your experiment logbook. "Tonight, we'll test fat ratios in pie crust. Tomorrow we'll study marination times for chicken thighs. Friday is pasta. Let's compare salted vs. unsalted water."

You try, record, taste, and iterate. Failures aren't disasters; they're data. Burnt cookies? You just overshot your heat curve. Flat bread? Your yeast underperformed. Bland soup? You skipped the acid adjustment. Every mistake is a lesson in thermodynamics and chemistry, delivered straight to your taste buds.

That's the rallying cry: cook like a scientist. Test, tweak, repeat.

***

Families as Research Teams

The best part? You don't have to do this alone. Invite your family, kids, and friends. Add them to the research team.

Kids especially love kitchen science because it's tactile. They stir, knead, whisk, and taste, getting immediate feedback. Forget textbook lectures. Hand a child a whisk and show them cream turning into foam. That's protein denaturation. They'll never forget.

Make it dinner and a science fair:

Why does salt make cucumbers sweat?

Does searing steak at high heat change juiciness?

What happens if we make cookies with baking powder vs. baking soda?

Around the table, the discussion isn't "Was it good?" but "What happened and why?" Suddenly, dinner is peer-reviewed.

***

The Do's of Kitchen Science

Like any lab, the kitchen has rules: **frame every dish as an experiment.** Hypothesis: More salt improves flavor. Test: cook two pots of soup, one salted and one unsalted. Collect data by tasting.

Do embrace control groups. Don't change everything at once. Adjust one variable at a time, such as sugar levels, cooking temperature, or fermentation time. Otherwise, your data will be noisy.

Do record observations. A notebook on the counter isn't weird; it's how discoveries happen. Write down ratios, times, and outcomes. Future-you will thank you.

Do celebrate "failure." The burnt batch is the baseline. You know where not to go. Celebrate like a scientist. Failure means progress.

The Don'ts

Don't mythologize. Cooking myths are superstitions masquerading as wisdom. "Searing seals in juices"? Nope. "Alcohol all cooks off"? Nope. Respect chemistry, not folklore.

Don't fear the mess. Science is messy. So is cooking. That's why sponges and dishwashers exist.

Don't stop questioning. If a recipe says "boil for 15 minutes," ask why. What's happening during those 15 minutes? How does it change at 10 or 20? Curiosity is your sharpest knife.

***

Meal Planning as Experimental Design

A scientist doesn't just run experiments; they design them. A good research plan is a menu. A week's worth of dinners can be a week's worth of chemistry labs:

Monday: starch experiments. Cook rice with different water ratios and chart fluffiness vs. stickiness.

Tuesday: protein trials. Chicken breasts, brined vs. unbrined. Compare juiciness.

Wednesday: vegetable chemistry. Roast carrots at 300°F vs. 450°F. Track browning and sweetness.

Thursday: emulsion challenge. Make vinaigrette with mustard, without mustard, and with egg yolk. See which stabilizes best.

Friday: fermentation fun. Start a sourdough starter. Feed it daily. Observe bubbles and smells.

By Sunday, you're not just fed; you're smarter. You've also taught everyone at your table that dinner isn't an end product; it's a process.

The Table as a Lab Bench

Scientists publish papers. Cooks serve plates. Both await critique.

Family dinners become discussion groups. "The soup was bland" isn't a complaint; it's feedback. "Why was it bland?" becomes a hypothesis. Maybe salt, maybe acid, maybe aromatics. You test next time.

Imagine kids debating starch retrogradation in mashed potatoes, or spouses arguing about the thermodynamics of searing. That's not just dinner. That's culture as science.

The point isn't perfection. It's an inquiry. The table isn't just a place to eat; it's the lab bench where conclusions are shared, and ideas are born.

***

Challenges: Push the Envelope

Once you've mastered the basics, throw down challenges.

Can you bake bread without a recipe, relying solely on ratios and observation?

Can you replicate a favorite restaurant dish at home by reverse-engineering flavor and texture?

Can you invent a new family recipe by combining known culinary techniques, such as lemon juice for brightness, caramelized onions for depth, and MSG for umami?

Each challenge is a test of skill and curiosity. Each result, whether a success or a failure, adds to your collective knowledge.

***

Why This Matters

Science is often framed as abstract, distant, and clinical, but the kitchen brings it home. Food is chemistry you can smell, physics you can chew, and biology you can ferment.

When kids grow up cooking as science, they don't just learn to eat; they also learn to appreciate the process. They learn to ask questions, design experiments, and analyze results. They learn skepticism, patience, and resilience. They learn joy.

The kitchen becomes a training ground not just for meals but also for minds.

Toward a Culture of Kitchen Scientists

Imagine if every household treated the kitchen like a lab. Not sterile or intimidating, but joyful and edible. Imagine if meal planning were lesson planning, family dinners were colloquia, and recipes were protocols to be tested, not commandments to be obeyed.

That's the culture we build. A culture where science isn't something done far away in white coats but something lived daily with spoons, pans, and laughter.

***

Conclusion: The Rallying Cry

Here it is, distilled: Food is science. Cooking is experimentation. Eating is data collection. Family is your research team.

Stop cooking passively. Start experimenting actively. Take notes. Debate hypotheses. Fail gloriously. Taste constantly. Teach children that "why" belongs in the kitchen as much as "what's for dinner."

Because, in the end, cooking is the longest-running experiment in human history. It's older than metallurgy, older than writing, and older than civilization itself. Every bite is a data point stretching back millennia. When you step into your kitchen, you join that grand continuum.

So grab your whisk. Fire up the burners. Pull out a notebook. And invite your family to the table, not just as eaters but as co-investigators.

Dinner isn't just dinner. It's science; delicious, messy, glorious science.

Run the experiment. Taste the results. Share the data. Then prepare for round two.

Epilogue – Leftovers

If you've read this far, congratulations. You're officially a food scientist. Not because you memorized reactions or built a sourdough starter that could qualify for citizenship, but because you started asking questions. Why does this work? What happens if I change it? What's really going on in the pan?

That's the mindset of science, and it doesn't end when you put this book down. Every meal is another chance to test, tweak, laugh, and learn. Some dishes will be triumphs. Some will be disasters. All of them will be data.

And that's the secret the experts don't say out loud: cooking isn't about recipes. It's about curiosity, persistence, and the joy of discovery, served hot, preferably with seconds.

So clear the table, grab a notebook, and start over. The experiment isn't over. It never is.